Ice Age Boundary Zone

Science Exploration by Rolf A. F. Witzsche

© Text Copyright Rolf A. F. Witzsche 2018
all rights reserved

The book contains the transcript and images of the science exploration video by Rolf A. F. Witzsche, with the above title at: http://www.ice-age-ahead-iaa.ca-- The book is a part of the transcripts series.

Lead in:

Global Warming (by the Sun) ended in 1998. The Sun is gradually 'dying' towards the full Ice Age. Solar activity has already collapsed to half since 1998, measured in Berillium-10 and Carbon-14 isotopes ratios, as well as in reduced sunspot numbers, reduced solar-wind pressure, and in reduced radio solar flux intensity. All the measurements that have been conducted that measure solar activity, have collapsed to half. The measured solar collapse is correspondingly reflected in the Earth getting colder year after year. The full Ice Age is near. We are in the boundary zone to it. Crop failures are increasing in many parts of the world and are getting increasingly more-severe.

To the degree to which the nations lose their food resources, the nations cease to function as viable nations, and ultimately cease to exist. When food can no longer be produced, the populations become refugees. While in some cases the lost food production can be compensated with imports, for as long as those are available, ultimately the effected populations find themselves evicted by the cold and disabled agriculture, with no place to go to.

Fortunately, the looming tragedy can be prevented in a highly developed human world, with the building of a technological New World that the Ice Age glaciation cannot touch, where agriculture can continue, and our food production with it. Tragically, nothing is moving on this front as if food production was of no concern to anyone. We should see massive movements happening in this front, and those moving fast. Instead, we see no movement at all. Thus, the gravest crisis in history unfolds before us.

A collapse process has begun by which human living in Canada, Europe, Russia, and large parts of the USA and China, is becoming increasingly precarious. The question needs to be asked here whether or not this trend to zero in human living will be reversed before the entire planet becomes largely uninhabitable in the 2050s by extreme cold and extreme lack of precipitation. That's the danger. The evidence is plain. The transition to the next 90,000-year glaciation phase is already in progress and is far advanced. One only needs to open one's eyes to see it. The universe is changing. One needs to move with it to live.

While the critical recognition of what drives the solar dynamics in the real world is blocked by a tragedy in science, a society of human beings still has the power to step itself past the blockage that shrouds reality, and discover the truth. The fringe effects that are already felt worldwide are getting more intense, while we have 30 years still remaining with increasing effects still to come with a potential magnitude that no one has experienced before. Thus we need to get serious and quickly on this front, while we still can, and start to build us a New World in which we can live, which also promises to become a renaissance world by the dynamics involved.

A new type of war is needed on this front - not a war to destroy our world and kill one another - but a great war against small-minded perceptions, to build up a strategic defense of humanity against the Ice Age consequences. That's what is required for humanity to have a future, even a brighter future with a richer civilization than we presently have, on an otherwise largely uninhabitable Earth.

Some indexing has been attached to the individual pages that provides an overview of the Ice Age Challenge and its many aspects.

Contents

- ➢ **The Ice Age is Coming** 11
- **Glaciation period is 500 years overdue** 12
- **Space exploration opened a new window** 14
- **Orbital cycles theory** 15
- **Mechanistic Ice Ages are not physically possible** 16
- **Ice Ages are caused by the Sun** 17
- **Berillium-10, that is generated by the Sun** 18
- **Scientific ignorance endangers humanity** 19
- ➢ **Facing the Ice Age Consequences** 20
- **Boulders oddly out of place** 21
- **Little Ice Age weak solar activity** 22
- **To measure the past in retrospect** 23
- **Who can fathom a 70% weaker Sun?** 24
- **80% less precipitation** 25
  - Water Crisis: Healing the Colorado River 26
  - California Drought - Science & Technologies 26
- **Full glaciation as near as the 2050s** 27
- **Solar-wind pressure began to diminish** 28
- **The solar wind will stop in the 2030s** 29
- **Ice Planet Earth with a hibernating Sun** 31
- **Interglacial high-powered Sun** 32
  - The Greatest Science Challenge in the History of Civilization 33
- **Then, 3000 years ago** 34
- **Measured in Carbon-14 isotope ratios** 36
- **The up-ramping of the Sun** 37
- **Over the hump of the global warming uplift** 38

Half-way along to the zero point 39

The Sun falls back to its hibernation mode 40

The phase shift is inevitable 41

Timing is determined by the threshold level 43

The rapid weakening that we see 45

These things are happening now 47

When the Earth becomes uninhabitable 48

    Failure is not and option: Dynamics of Wisdom in an Ice Age World 48

A crisis in history unfolds at the present stage 50

The giant is asleep 51

A city downstream from a great dam 52

The Ice Age is close to be breaking 53

➢   Humanity is riding the 'Merry Go Round' 54

My numerous science-video productions 55

The metaphor of the carousel 56

The Ice Age Challenge a bugle call to us 57

➢   What do we know about the Ice Age dynamics? 58

Most people are fast asleep 59

In the ice core record 60

Immense cold under glaciation conditions 61

The up-ramping of the Sun in 1715 62

Without it, none of us would exist 63

The world population of the last Ice Age 64

Now, we are 7 billion people 65

The entire development of civilization 66

➢   We need to build us a new world in the tropics 67

We need to build 6,000 new cities 68

- ➢ A Wellsian Crisis Project: To disable science 69
- To stop the march of science by all means 70
- H. G. Wells wrote the script for the process 71
- A debate as to how to deal with science 72
- The Hydrogen-Fusion Sun doctrine was invented 73
- And a bit later, the Big Bang doctrine 74
- The 'Invariable Solar Constant' 75
- The 'devils fork' came out of the 1920s 76
- If the Ice Age 'cat' was to slip out 77
- Ice Age recognition kept out of sight 78
- ➢ What forces cause an Ice Age? 79
- Recognition of the Sun as a sphere of plasma 80
- All the known parameters about the Sun 81
- A Plasma Sun is an extremely variable star 82
- Since the Sun is externally powered 83
- A sink effect is needed 84
- The Plasma Sun is largely hollow inside 85
- In extremely large spheres of plasma 86
- A plasma sphere has the greatest mass density at its surface 87
- The pinch effect 88
- High-energy discharge experiments 89
- Toroidial structures are visible 90
- In the case of the Sun 91
- The heliospheric current sheet is aligned 92
- In the case of the Sun 93
- In large plasma streams, the Primer Fields 94
- In principle the powering of a Sun 95

Experiments, by LaPoint and Peratt 96

NASA's Ulysses satellite that had orbited the Sun 97

Verification in cosmic space 98

➤ **To 'see' the invisible** 99

Ulysses, had made the invisible, 'visible.' 100

Another 'visible' example of plasma streams 101

Shape being that of a confinement dome 102

The galactic node point 103

In a hypothetical case for our galaxy 104

We have made it visible 105

Galaxy is presently at the weakest point 106

Progressively more severe 107

During the last half-a-million years 108

Only 15% of the interstellar resonance cycle 109

During those weak times 110

Global cooling had exceeded 40-fold 111

Cooling reflects a 70% weaker Sun 112

The vast majority of the stars 113

Our Sun, as a rather mediocre star 114

➤ **Seen the interglacial climate diminishing** 115

No one in general society is aware 116

We are the 8th human species 117

The most critical juncture in our entire history 118

We have 'seen' in Beryllium records 119

We have seen the Sun getting weaker 120

Solar warming events diminishing 121

Solar minimum events diminishing 122

We now see the solar cycles diminishing ... 123

We even see the solar-wind pressure collapsing ... 124

We now see the solar weakening progressing ... 125

We have watched all of these events unfolding ... 126

Project the consequences into the future ... 127

We have made enormous scientific efforts ... 128

As near as 30 years from now ... 129

The near future is a death-trap without ... 130

We remain stuck in our beloved easy chair ... 131

➢ Getting out of this trap ... 132

The Sun as its own master ... 133

Let the magic tales go ... 134

Magic tale of the universe exploded ... 135

The real universe is a plasma universe ... 136

The Big-Bang red-shift a deception ... 137

A new world comes into view ... 138

The Plasma-Sun is actually physically possible ... 139

The Gas-Fusion-Sun concept ... 140

The Plasma-Sun brings the solar system into the universe ... 141

The gas-sun is the most deadly doctrine ... 142

So, who is afraid of the Ice Age? ... 143

To be part of the Ice Age Renaissance world ... 144

Just imagine producing 6,000 new cities ... 145

➢ The power of the universe is aiding us ... 146

Increase in cosmic-ray events beneficial ... 147

Cosmic-ray particles 'see' our biology as empty ... 148

Electric currents by magnetic induction ... 149

Electric currents beneficial 150

The greatest cultural advances 151

Lean times of cosmic-ray flux, times of war 152

With the solar system getting weaker 153

The age of the wars appears to be over 154

The greatest challenge in the history of humanity 155

➢ More from the author: 156

14 Libraries of books and video productions.. 156

➢ The Ice Age is Coming

Wake up! The Ice Age is Coming! The Ice Age is coming! Get the fur coats out of the closet.

The above scene is from my garden (app 2010). It is featured in my first major Ice Age exploration video with the title Ice Age Precursors, produced in 2011 - a 6-part video that covers a wide range of topics, from cosmic dynamics to politics. (The video opens symbolically with the sound of silence - a world without a human voice).

The "precursors" video is twinned with an exploration video of how humanity may be able to break out of the trap of its indifference to the Ice Age Challenge. This is still the greatest challenge in civilization, to protect civilization, nourish human existence, and to have a future. The historic video is produced in two parts: Ice Age Breakout 1 and Ice Age Breakout 2.

Glaciation period is 500 years overdue

That's the message that the world renowned atmospheric scientist, the Polish Professor Zbigniew Jaworowski, chairman of the Scientific Council of the Central Laboratory for Radiological Protection in Warsaw, presented to the world in his 2003 paper with the title "The Ice Age is Coming".

He warned that the transition to the next glaciation period is 500 years overdue, and can start quickly without warning, and might be as short as 1 to 2 years.

He couldn't say more, because the forces that are driving the Ice Age dynamics had not been discovered at his time in 2003. The scientific foundation for making the discoveries hadn't been established.

And even now, the discoveries that were made are largely ignored. The world is living in the dark in this respect.

The prevailing theories at the time, of what causes the Ice Ages, were so far off the mark that Jaworowski didn't even bother to refer to them in his paper on the Ice Age.

- The Ice Age Is Coming! (pdf)
  21st Century Science and Technology
  - Winter 2003-2004 Zbigniew Jaworowski

  Prof. Dr. Jaworowski refutes the Manmade Global Warming myth in this article. He states that CO_2 is not a climate factor, so that the doubling the atmospheric the atmospheric CO_2, if it was magically possible, would have a trifling effect, even the burning of all carbon resources that we can lay our hands on.

And so the world keeps on dreaming of global warming, while, by its indifference to the physical reality, it lets its future slip away from under it by not preparing it for the coming Ice Age.

Se my video: Manmade Global Warming Impossible

Space exploration opened a new window

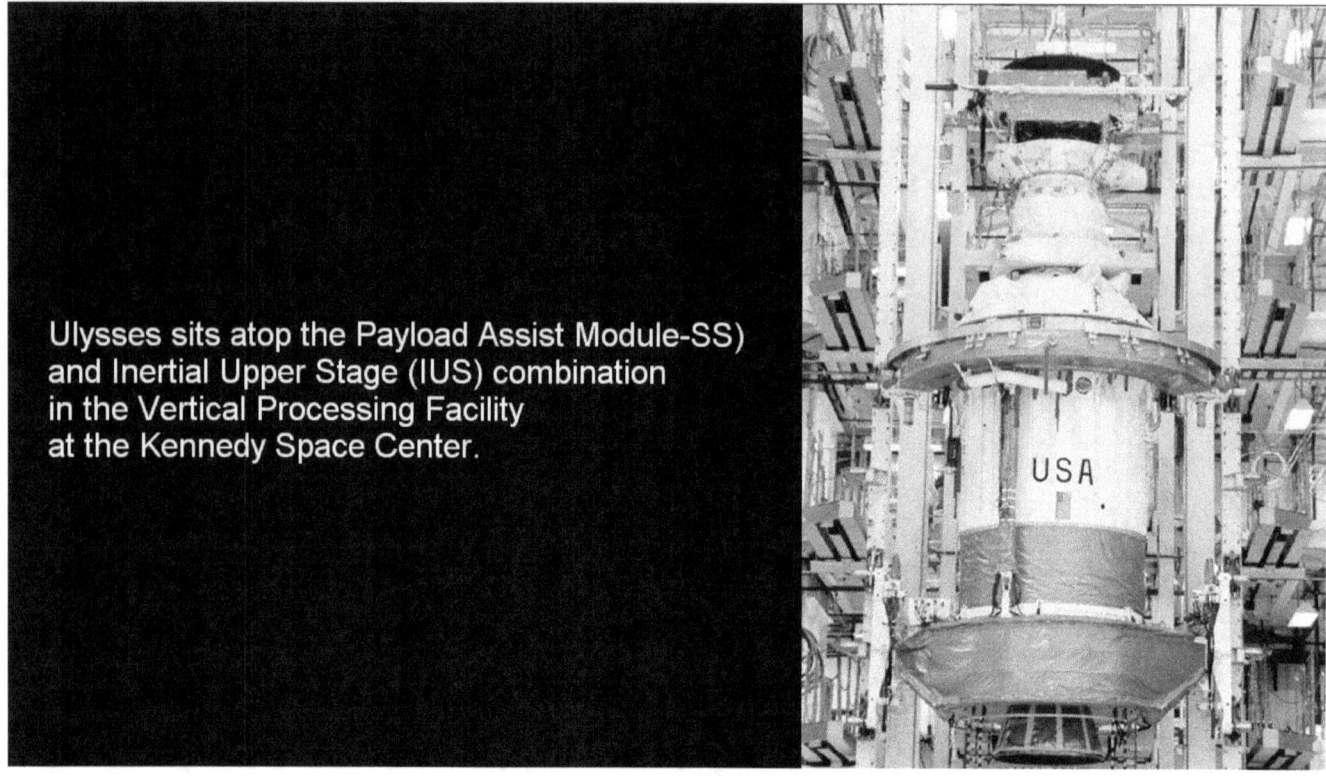

Ulysses sits atop the Payload Assist Module-SS) and Inertial Upper Stage (IUS) combination in the Vertical Processing Facility at the Kennedy Space Center.

Until the early 2000s, when space exploration opened a new window to the universe, the prevailing Ice Age theories had been guesswork.

> NASA's Ulysses mission opened up a new horizon for exploring climate dynamics in the solar system. It provided the first in-space measured evidence of the weakening of the solar system in a near free-fall collapse fashion.
>
> See: Ice Age Start-Up Phase I 45% Complete

Orbital cycles theory

The most prominent of these was the orbital cycles theory that had been developed in the 1920s, nearly 100 years ago.

The mathematician Milutin Milankovitch, had theorized at the time, for the lack of real data, or for other constraints, that Ice Ages might be caused by the combined effect of variations in the eccentricity, axial tilt, and precession of the Earth's orbit around the Sun, which he theorized would result in cyclical variation of solar radiation affecting the Earth.

While he failed to realize - that the total energy that is received on the Earth is always the same, regardless of the orientation of the Earth's spin-axis, which affects only hemispheric distribution; and he also ignored what Johannes Kepler had already stated in the 1600s, that the changing eccentricity of the Earth orbit abound the Sun, doesn't affect the total solar energy received in any way, but affects only the seasonal distribution - Milankovitch's theory was hailed nevertheless, and became the mainstream theory.

> The Milankovitch theory is still being hailed, even while it has been demonstrated in ice core records that the computed results and historic climate pattern don't match. Sometimes they come close, as one would expect from the evident fact that the orbital variations are subsequent phenomena of the Ice Age dynamics and are not causative for them.
>
> See: New Ice Age Near: 58-part Evidence

Mechanistic Ice Ages are not physically possible

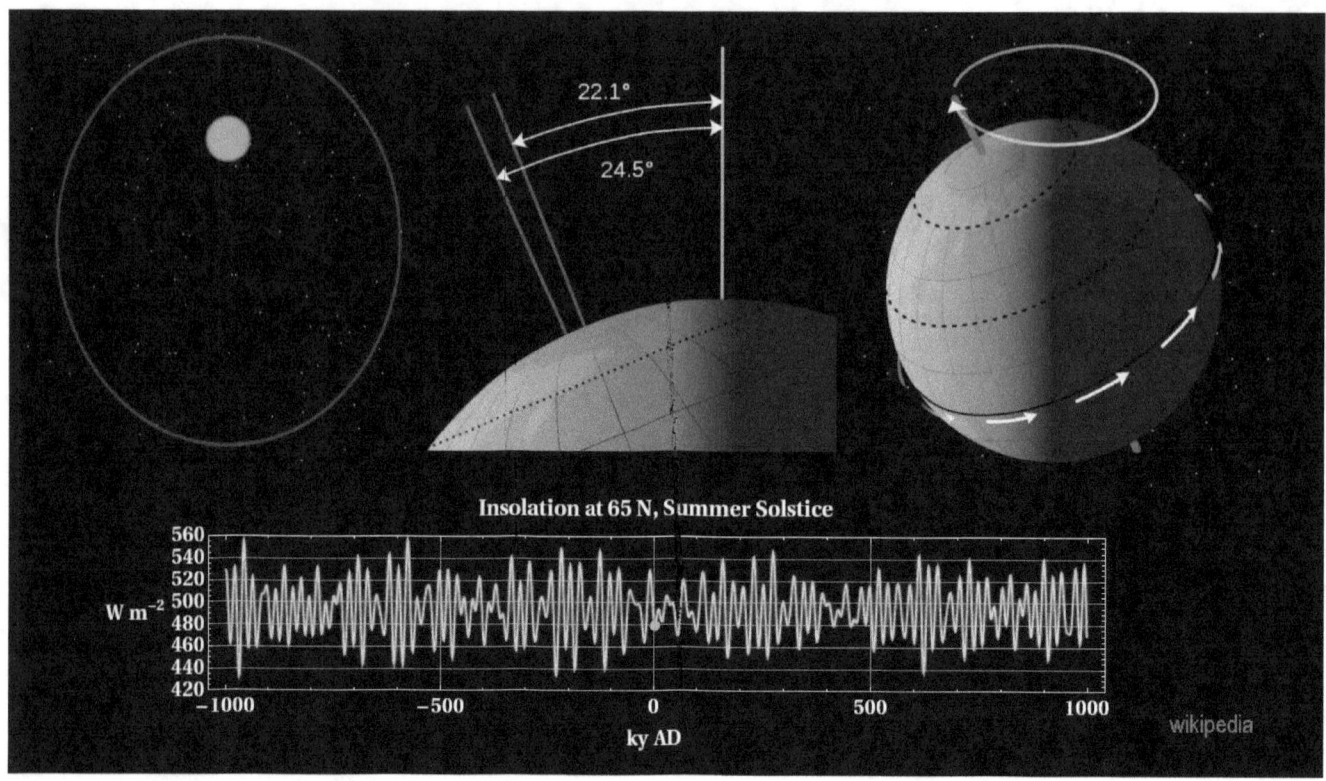

The fact is, that astrophysical science has been latching onto straws for a very long time, in attempts to rationalize the occurrence of Ice Ages as mechanistic phenomena, which they are not, because mechanistic Ice Ages are not physically possible.

> Since the orbital variations only affect changes seasonal and hemispheric distribution, it is nor surprising that the computed results don't match historic results, since the over-all effect is too minuscule to matter.
>
> See Ice Age Precursors transcripts, Part 2, Dynamics vs. Mechanistics

Ice Ages are caused by the Sun

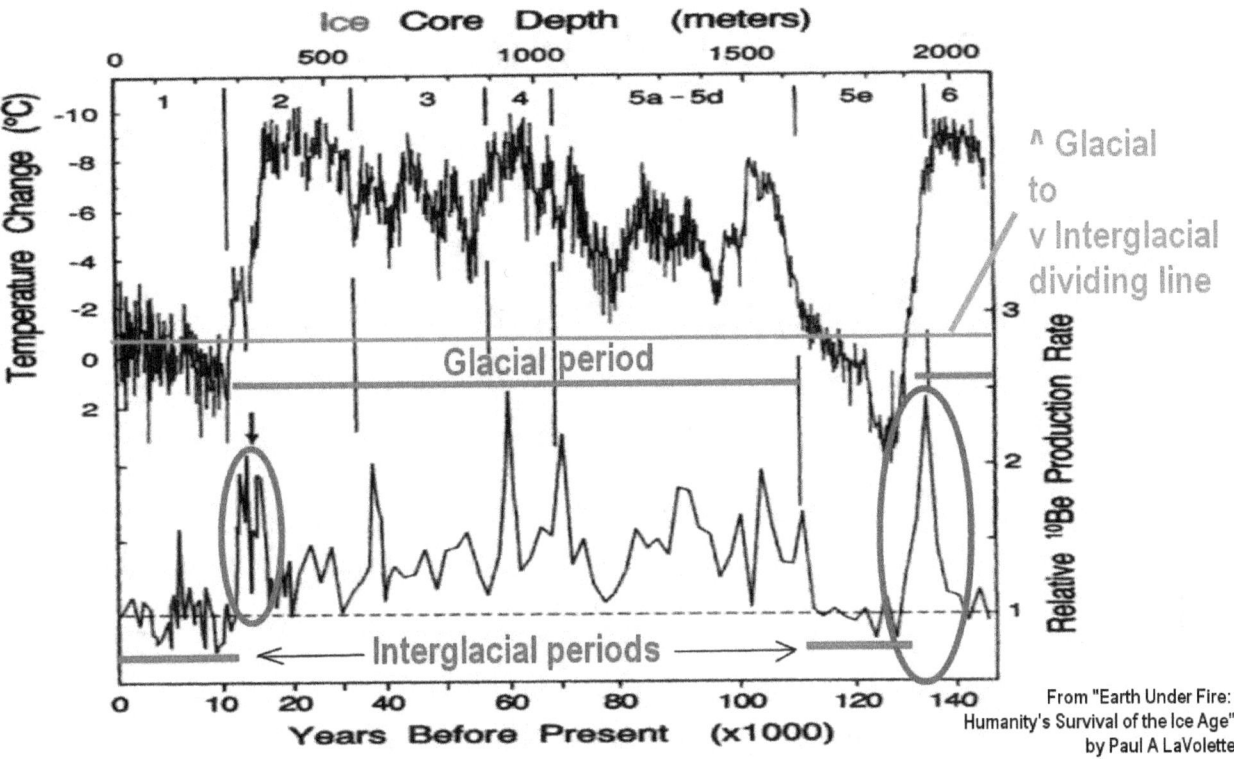

Now, 100 years later, thanks to modern technology, we have physically measured evidence on hand that proves conclusively that Ice Ages are caused by the Sun directly and exclusively. No guesswork enters the scene. Orbital cycles, ocean current fluctuations, and CO2, none play a role. Ice Ages are caused by the Sun. The evidence is hard and conclusive, though it is typically ignored in the science communities.

> The Sun has provided us a proxy for its historic activity intensity in the form of the Beryllium-10 isotope. The result is extensively explored in the video: Grand Solar Minimum becomes the Ice Age

Berillium-10, that is generated by the Sun

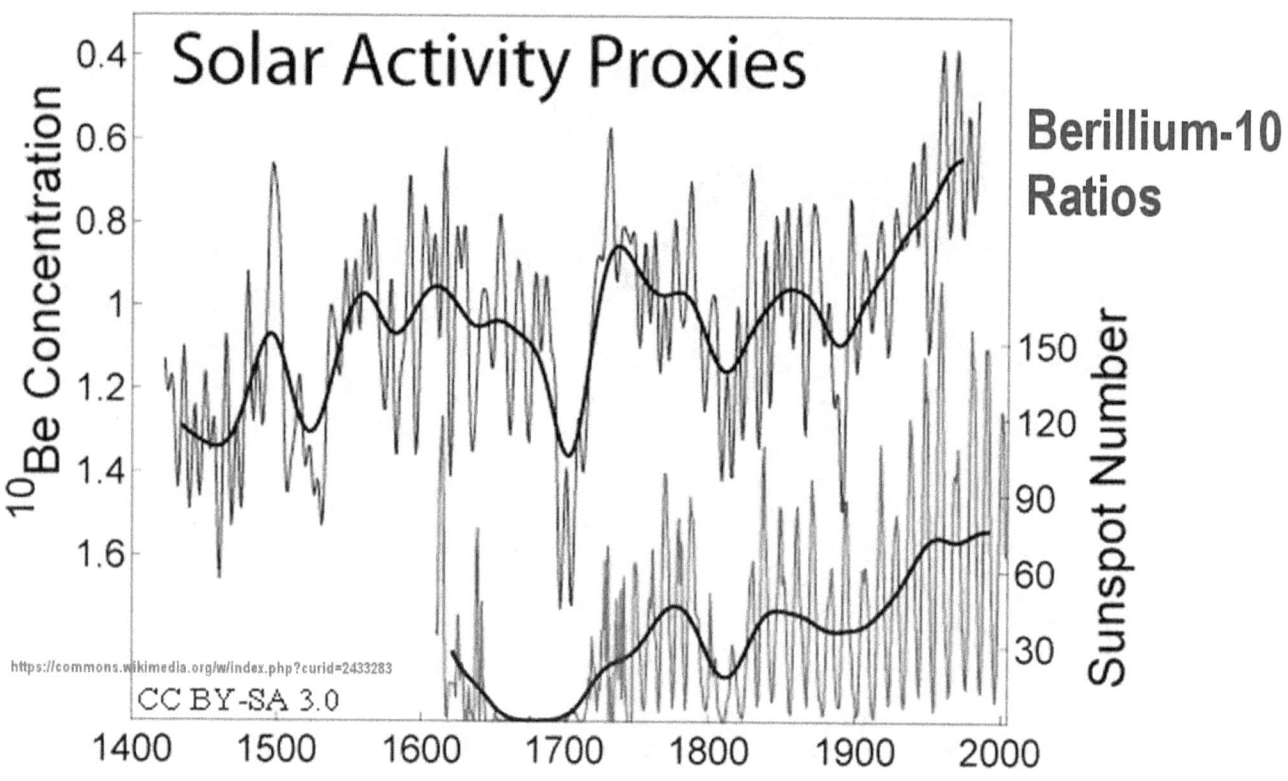

The evidence that is laid up in ice, exists in the form of changing production rates of the isotope Berillium-10, that is generated on the Earth by the Sun. The Beryllium-10 production rate can be measured. It varies in lockstep with solar activity. This means that historic Beryllium-10 ratios can be measured as a proxy for historic solar activity.

> We have 3 measurable proxies available, for historic solar activity intensity. One is the sunspot numbers. High numbers represent high rates of solar activity. Another is the production rate of the radioactive isotope Beryllium-10. It varies with solar cosmic-ray flux. High values represent a weak Sun the generates large volumes of cosmic ray events, which when they interact with the Earth's atmosphere generate both the Beryllium-10 isotope and also the Carbon-14 isotope. We also know from historic experiences that low sunspot numbers, which represent low solar activity, also become manifest on Earth as cold climates.
>
> (see item 17-22 of transcript Part 1)
> of Ice Age Start-up: 1st phase now 45% complete

Scientific ignorance endangers humanity

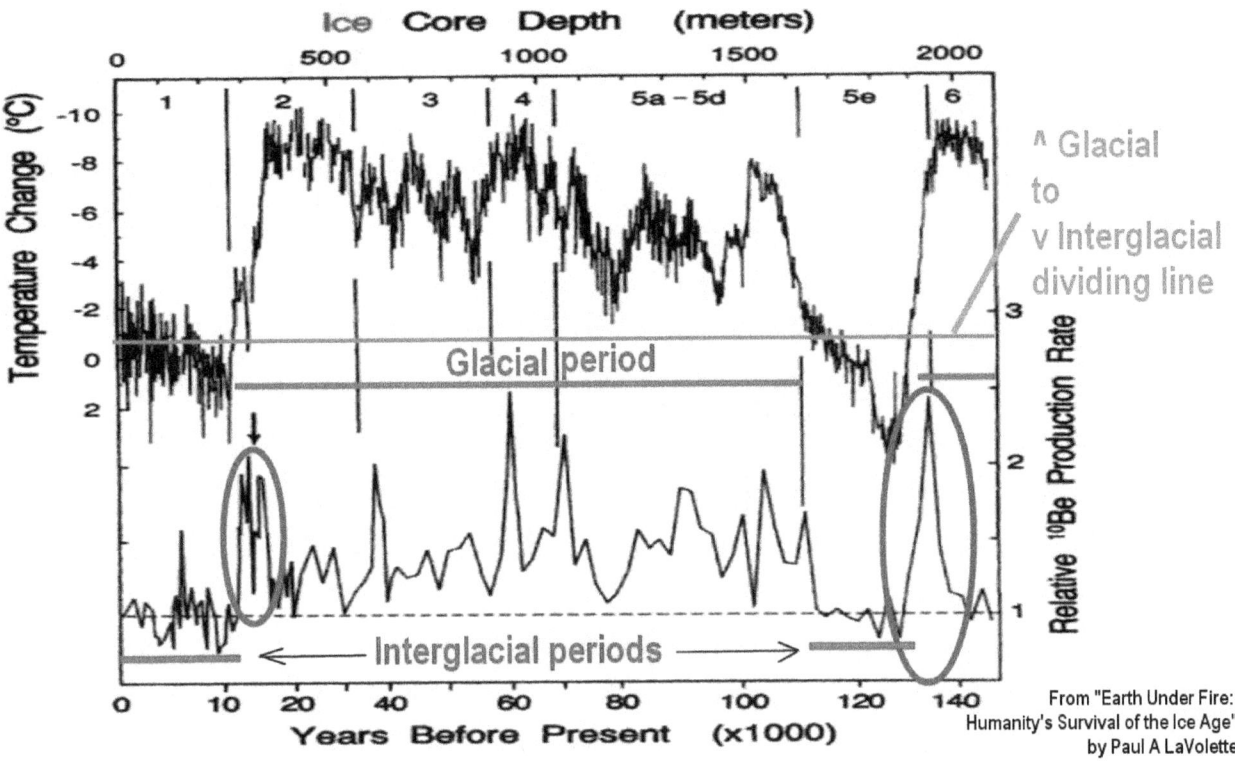

From "Earth Under Fire: Humanity's Survival of the Ice Age" by Paul A LaVolette

The consistently high Beryllium production rate throughout the previous glacial period, but not before or after, proves that the previous glacial period - termed the Ice Age - was not caused by orbital mechanistic, but was caused by the Sun directly, so that we can say with scientific certainty that the Sun causes the Ice Ages, and nothing else does.

While working in the dark, without a clue of what is actually happening, by ignoring the measured physical evidence, some scientists still predict that the next Ice Age won't start for another 50,000 years. Some even say that manmade global warming will overpower the Ice Age dynamics, as if we had the power to affect the Sun.

With these types of denials of reality, the scientific community has put itself to sleep, and humanity with it, almost universally. And more than that, the scientific ignorance endangers the whole of humanity, as it thereby prevents society from building itself the technological infrastructures that are needed for continued human existence during the Ice Planet phase of the Earth.

See: The Grand Solar Minimum becomes the Ice Age

➤ Facing the Ice Age Consequences

> Facing the Ice Age Consequences

Facing the Ice Age Consequences

Boulders oddly out of place

A story is told, that back in the 1800s, a traveller in the Alps had noted giant boulders in a valley. The boulders seemed oddly out of place. The traveller was puzzled by how they got there. When he questioned nearby villagers, he was told to look at the distant glacier high in the mountains. They told him that the glaciers had once extended down into the valley during the cold times, and had left the boulders behind when they melted."

With this 'discovery' an interest was roused in large-scale of glaciation phenomena that can have such amazing effects.

> For more background details see item 28-49 of transcript Part 1 of Ice Age 2050s: 'Certainty'

Little Ice Age weak solar activity

The glaciation period that the villagers spoke of, had evidently been that of the Little Ice Age of the 1600s. We know today that the Little Ice Age had been a period of extremely weak solar activity. We also know that the climate had been so cold in those days, that rivers became skating rinks, and agriculture diminished so extensively that between 10% to 30% of the population in Europe had perished by starvation, depending on location.

This harsh period of cold, which had spanned more than a century, had nevertheless been but a fringe effect on the path to the big Ice Age, the deep glaciation period that typically spans 90,000 years. No one in the 1600s could have imagined what a full Ice Age has in store for humanity, which would have paled the harshness at the time into insignificance.

We came close to loosing the interglacial climate at the time. If the Sun had not been ramped up in 1715, the transition to the Ice Age might have occurred in the 1700, during the then ongoing diminishment of the Sun.

See: item 24 of transcript for Ice Age 6-in-1 Flank

To measure the past in retrospect

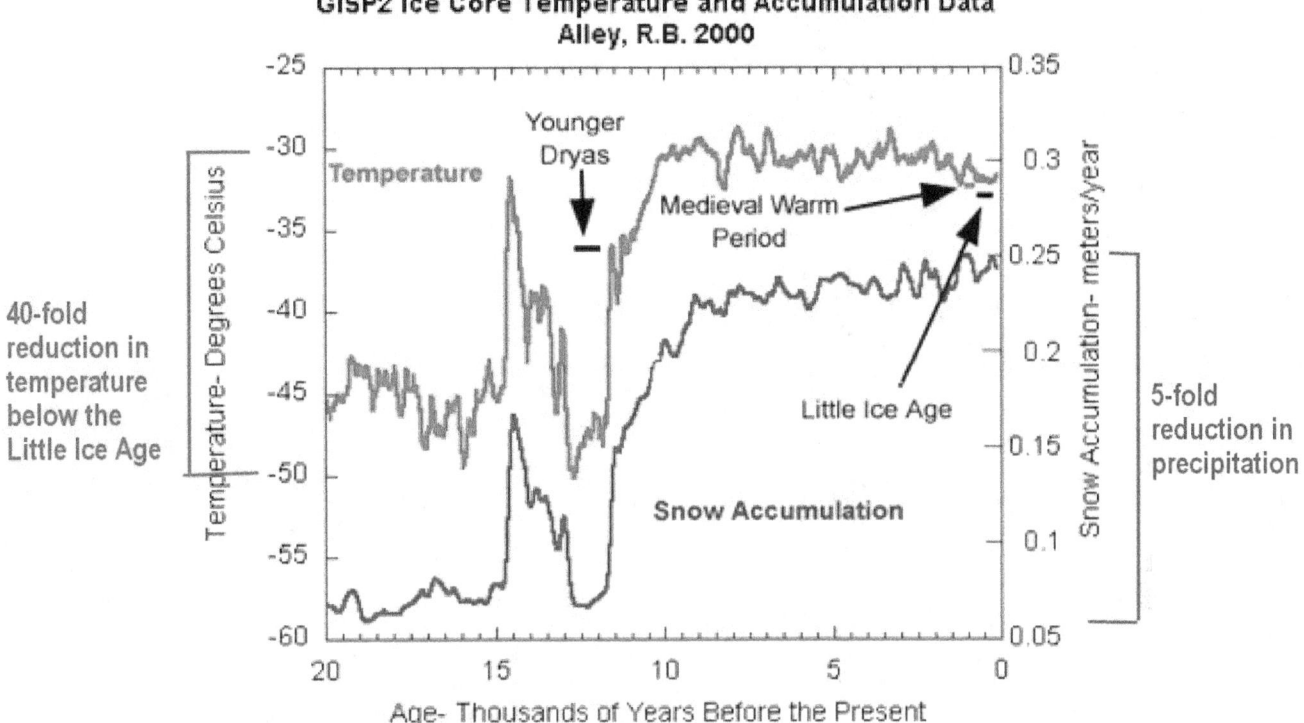

We know a little more today along this line. By analyzing the content of the ice of the giant ice sheet of Greenland, it became scientifically possible to measure the past in retrospect, and to reach far back in time.

It became apparent in ice core records that the cooling of the Earth, during the last glaciation period, had been 40 times more extensive than the Little Ice Age cooling had been, which had extended high mountain glaciers into a Swiss valley. It also became apparent in the ice core records that the precipitation on Earth had been 80% less during glaciation time, than it is today.

What you see here are real physical measurements, carefully plotted, meticulously produced. The numbers are big.

See items 1 - 15 of transcript Part 1 of Ice Age is Digital - Awake

Who can fathom a 70% weaker Sun?

70% weaker Sun during glaciation

It is almost impossible to imagine what those big numbers are telling us. Who can fathom, for example, the consequences of a 70% weaker Sun? For starters, the liveable zone shrinks to the latitudes of the tropics.

> We won't likely regress to the glaciation stage with a 70% weaker Sun for yet another 30 years. Till this phase shift happens, we will remain in the interglacial 'warm' climate. However, this climate is now rapidly diminishing to ever colder and dryer climates and will continue to get colder and drier for another 30 years.
>
> The fringe effects are already severely affecting the present world with untimely and extreme climates that result in crop failures, empty grocery shelves. Since this is the beginning of a 30 years trend, we will soon see worldwide food shortages, and mass migration to avoid starvation. We are presently close to the Dalton Minimum level, in solar cycle 24. When we reach the Maunder Minimum level thereafter, potentially in the mid 2020s, we can expect to see hunger on a scale not seen before. We may see borders being closed to prevent mass migration.
>
> The tragedy can be still avoided by massive developments of new agriculture, and water management systems. See: Freshwater Unlimited

80% less precipitation

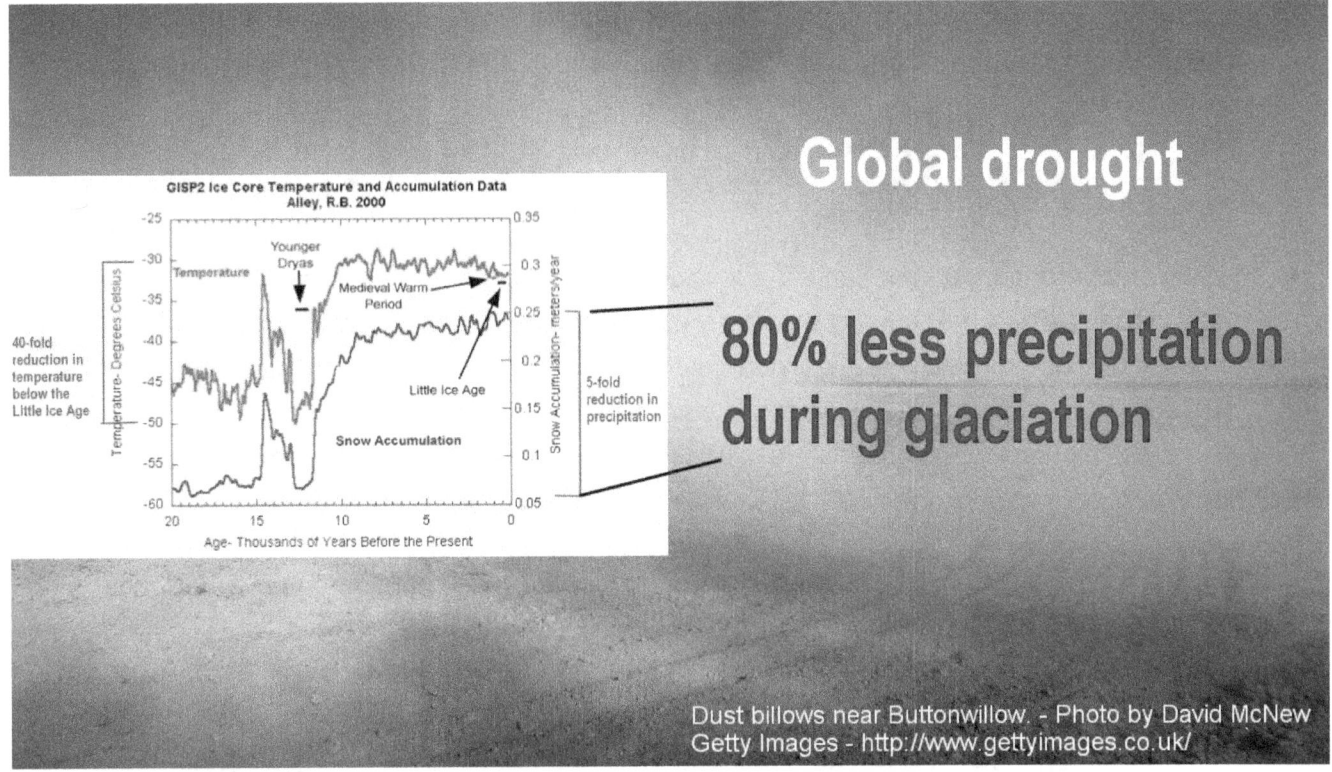

80% less precipitation during glaciation

Global drought

And the 80% loss of precipitation means that nearly the entire zone that remains liveable on our planet, becomes a desert.

That's the type of situation that we will face from the 2050s onward. That is what we need to get ready for. The liveable zone is the zone that can be made liveable with large-scale technological infrastructures, which can be built extensive enough to accommodate 7 billion people. This means that we need to build us a completely new world in which the Ice Age cannot affect us, and that we get this completed in the 30 years that we have remaining of the interglacial climate.

> While we still live in the interglacial climate, severe drought conditions have already begun. Two major fringe effects with the potential to disrupt out food supply, are already experienced as the result of the weakening solar system. One is flooding; the other is the drought. Historic experiences indelicate that the first effect of the weakening Sun is felt in the form of flooding, which eventually gives way to evermore expanding drought conditions. This is evident in flood level markers in a city on the Donau River in Germany. During the early stages of the Little Ice Age in the 1500s, the Donau flood levels were marked high up on a wall. This high flood level reflects high volumes of solar cosmic-ray flux. But during the coldest part of the

Little Ice Age, at the Maunder Minimum of solar activity, no flooding occurred. Instead, tree rings examined for this period indicate that a severe drought had gripped the Earth, lasing several decades. This is what we will face in the late 2020s, most likely, when the currently weakening Sun falls back to the Maunder Minimum level.

See: {item 36 - 45 of the transcript Part 1, and 80 - 84 of Part 2) of <u>Cosmic Climate Change - Part 2 From Big, to Gigantic, to Joy</u>

While drought conditions can be eliminated with technological means, in a world that hails depopulation, nothing is being done to protect agriculture from drought condition. The governor of California responded to the big California drought from 2010 onward, saying "Let the people who don't like the drought, move away. But with extreme drought becoming world-wide in the late 2020s, the governor's advice, to simply move away, cannot be implemented if society has no place to go to.

NASA's Ulysses satellite gave us a view into the future. But nothing was done. Instead of building the infrastructures for humanity to have a future, society wasted its resources with war, destruction, and looting one another. Nevertheless, those rescuing infrastructures can yet be built, before the Maunder Minimum stage is reached the 2020s, and the mass starvation of humanity begins with no place to escape to.

See: <u>Water Crisis: Healing the Colorado River</u>

And <u>California Drought - Science & Technologies</u>

Full glaciation as near as the 2050s

It can be stated with a fairly high degree of certainty, based on events, measurements, and numerous forms of evidence, that the phase shift to the next period of full glaciation is potentially as near as the 2050s, if not closer, perhaps even as close as 30 years from now,

The crucial factor, of course, is the timing. The time-frame for the phase-shift to be happening in the 2050s wasn't pulled out of a hat. It is the composite result of numerous events and measurements that all point to the 2050s in their own unique way.

> Ice Ages are caused by the Sun. For this to be possible, the Sun needs to be a variable star. But how is this possible? The historic exploration series, Ice Age of the dimming Sun in 30 years - a series of 10 videos - is designed to present the numerous aspects of it.

Solar-wind pressure began to diminish

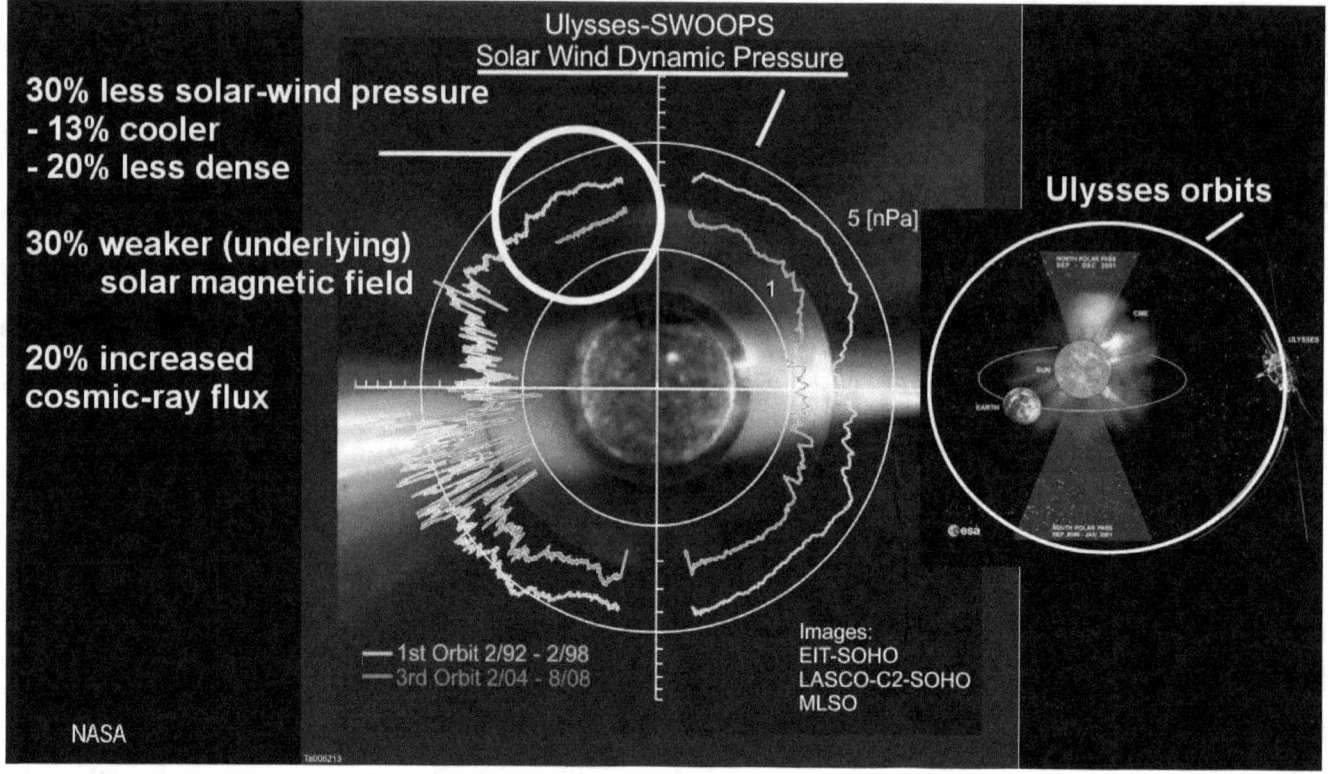

For example, as measured in space, the solar-wind pressure began to diminish from the year 2000 on, or slightly sooner, at a rate of 30% per decade, and likewise the solar magnetic field, while the solar cosmic-ray flux was increasing.

> NASA's Ulysses spacecraft gave us a measured rate of the collapse of the solar system by measuring the diminishment of the solar-wind pressure, monitored over the timeframe of a solar cycle. The measurement gives us a basis for determining the potential phase-shift timeframe for the start of the next Ice Age (potentially in the 2050s).
>
> See item 23-37 of transcript Part 1 of Ice Age Start-up: 1st phase now 45% complete

The solar wind will stop in the 2030s

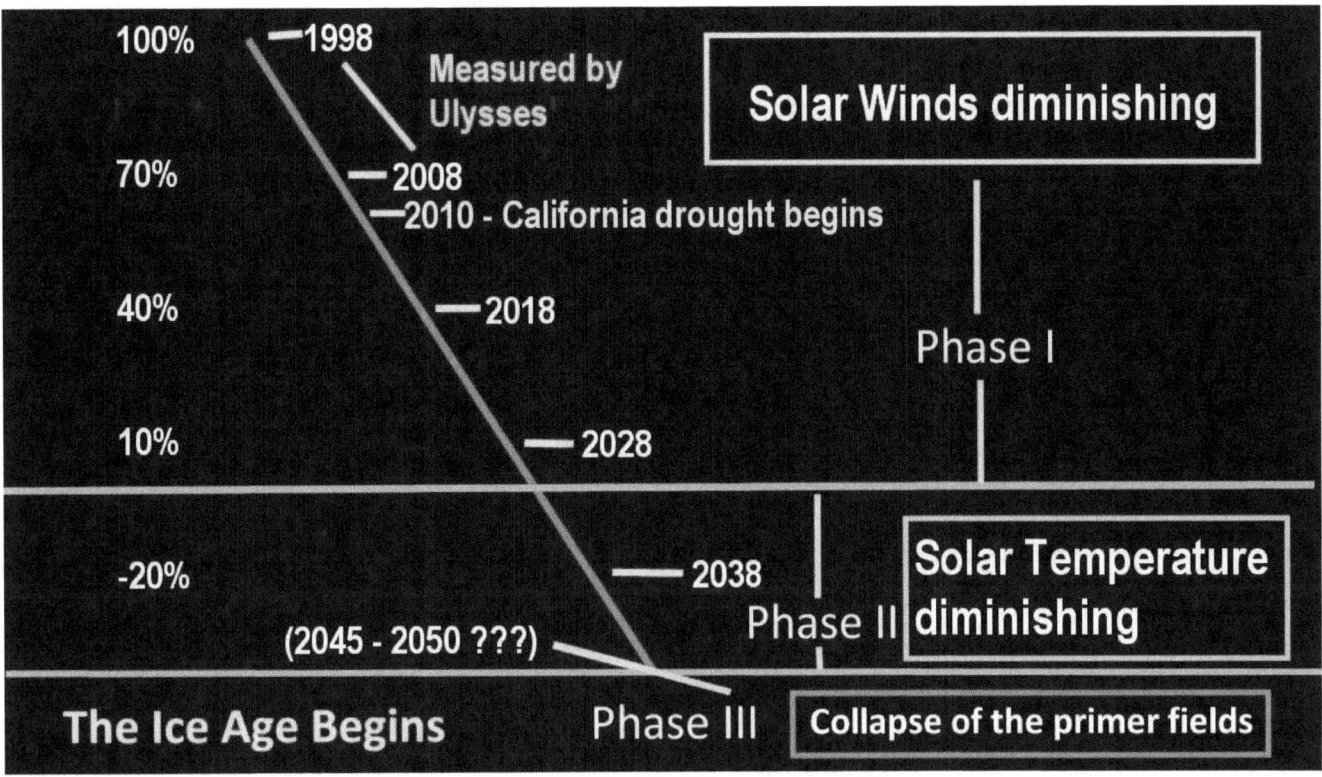

If one projects the measurements forward, the solar wind will stop in the 2030s, and the Sun's temperature will diminish thereafter till the Ice Age phase shift happens, with which the interglacial ends.

The measurement of the collapsing of the solar wind gives us the potential date of the coming 'exam' time, when the question becomes paramount, of whether or not the large-scale infrastructures have been built that enable human existence to continue on an Ice Planet Earth, that under present parameters becomes largely uninhabitable. If we make the grade, with the needed infrastructures in place, we pass the 'exam' and will be able to live. If not, 99% of humanity will perish by default.

However, there is a prior exam called, that we must pass, before we even get to the phase-shift event. If we don't build the infrastructures that enable our present agricultures to remain intact and productive under the worsening fringe effects, mass-starvation will destroy much of humanity. We knew from 2008 on, from the measurements provided by the Ulysses spacecraft that we would face this critical threshold, while absolutely nothing was done to qualify us for continued living. We wasted 10 years, by sitting idle.

Fortunately, we can still make the exam with a crash program to build the worldwide preparations. But will we do it? We have forced us into a rut by small-minded thinking, religious control, imperial domination, and economic insanity, so that nothing much works anymore. Even science has been decimated that way, to the point that physicists don't believe in physics anymore, with a few exceptions.

To date, China is the only country committed to large-scale science, infrastructure, and water development, with a special focus of developing Africa, which has the potential to become the food basket of the world, till the Ice Age Phase shift happens. China appears to be aware that a food crisis looms on the horizon of the near future by the weakening Sun and is preparing for it by developing Africa. and by encouraging worldwide economic cooperation on a wide front for infrastructure building.

See: (Transcript) of Peace and Joy and Power

Ice Planet Earth with a hibernating Sun

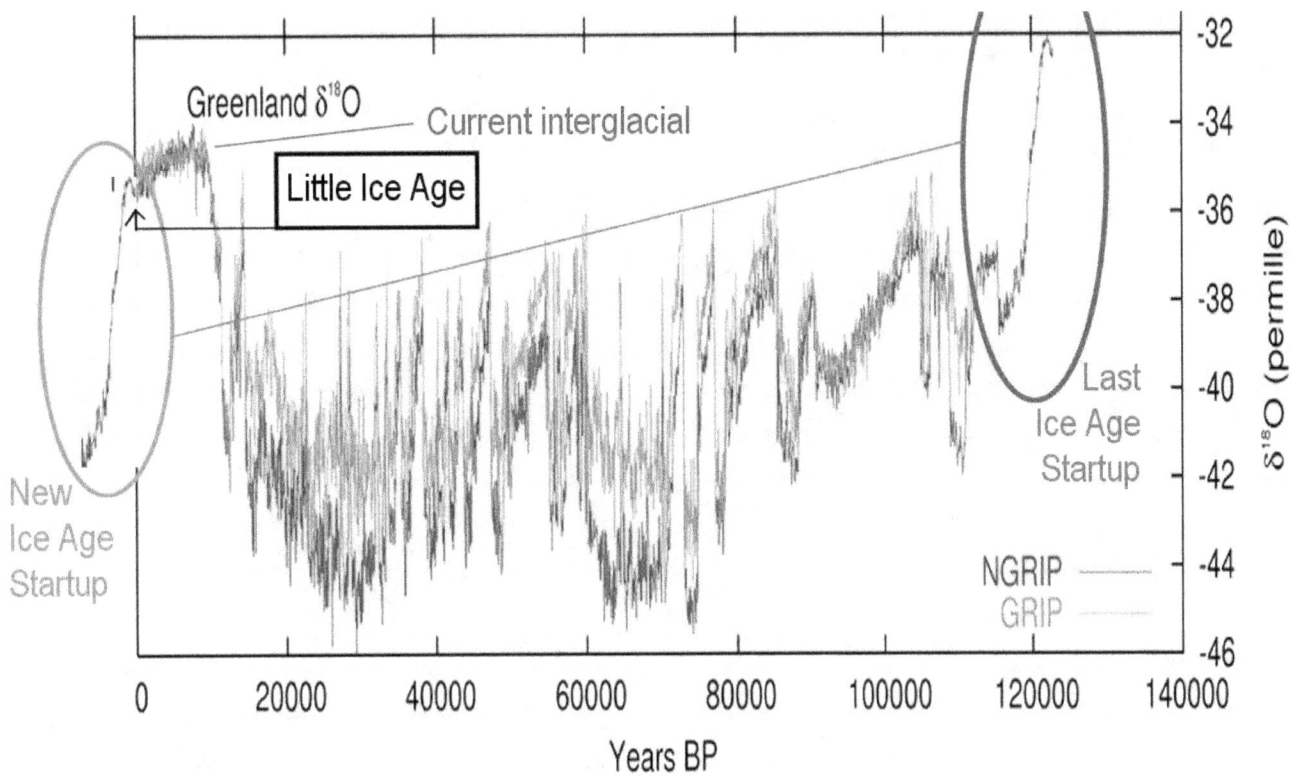

The Ice Age phase shift is the result of a threshold condition on the weakening slope of the solar system, beyond which begins a new cold world, a world of an Ice Planet Earth with a hibernating Sun. The phase shift causes a climate collapse as it did 120,000 years ago, when the previous Ice Age began.

If we built us the infrastructures in time, we can ride out the climate collapse and the 90,000-years-long Ice Planet (glaciation) phase. Two challenges stand in the way. One is our indifference to the challenge, and the other is our continuing commitment to war and even nuclear war. Our survival on these two fronts, depends on us gaining our humanity back that has been carelessly surrendered over many years for a long list of little and big issues. The project that we must master in order to have a chance to rebuild our humanity, is the Principle of Universal Love, a love for our common humanity that we all share. This may be the toughest project that we yet have to master, to create a path to where we can win and have a future. The exploration for this project unfolds in the form of my 14 novels designed for this purpose.

Interglacial high-powered Sun

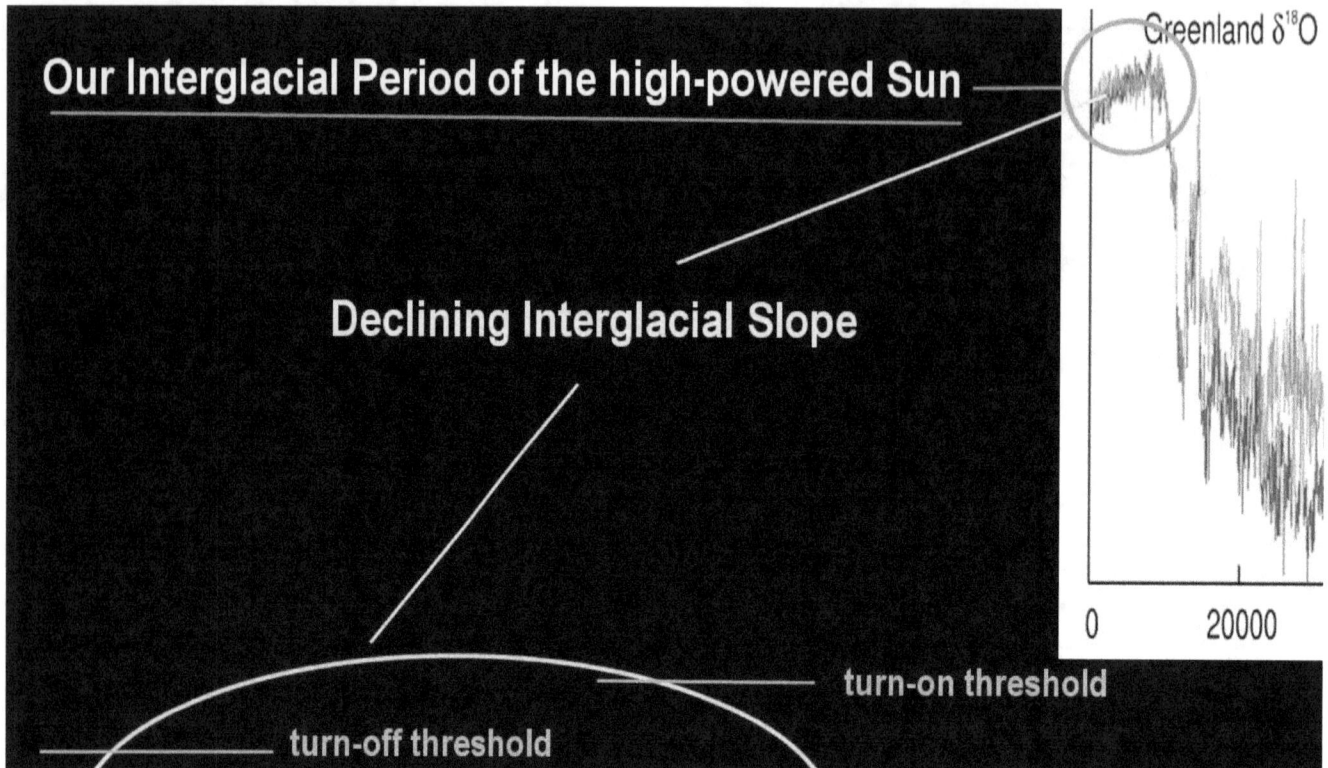

Our Interglacial Period of the high-powered Sun

The phase-shift away from the current interglacial period is determined by the turn-off level for the high-powered Sun. The turn-off occurs at a lower level than the turn-on. Greater plasma density is required to form the magnetic fields that focus dense plasma unto our Sun, than is needed to maintain them. The 'delayed' turnoff puts the turn-off point onto a steeper slope where the weakening of the Sun happens evermore rapidly before the turn-off.

We can tell that we get near the turn-off phase shift when the fringe effects increase dramatically.

The Fringe Effects of the Weakening Sun began 5000 years ago, when the interglacial climate began to weaken.

> Actually, the climate instability caused by the diminishing Sun, began already 7000 years ago. The climate has been on a roller coaster ride ever since, with a cold period 5000 years ago, followed by a recovery that peaked 3000 years ago and then dropped off to the present low level that is becoming evermore precarious. We at a point now where the increasing dynamics make historic comparisons evermore invalid. Historic records are laced with reports of famines and drought condition, but these have been mostly local. Now they are becoming global in scale. The weakening of the Sun has reduced the solar activity already close to Dalton Minimum level, which will drop off in the 2020s to the Maunder Minimum level, and evermore after that

during the 2030s and 2040s, perhaps even the 2050s, till the solar dynamics themselves collapse to a default level, and glaciation begins.

My point is that we are in a crisis state already, before the Ice Age even begins, where the fringe effects are becoming large beyond historic proportions. While large-scale crop failures are still localized phenomena, by local climate anomalies, humanity has never experienced world-wide simultaneous crop failures by global drought conditions. Nor is there an escape possible to avoid the consequences. There exists no place on the planet where society can migrate to in order to escape the drought. The only option we have is to develop large-scale artificial water infrastructures, and large scale protected agricultures, and to place those into the tropics as much as possible. The point is that the Ice Age isn't something that may affect us in the distant future 30 years from now, but is affecting us hard already, with devastating fringe effects, which hopefully will jolt us out of the easy chair to build us a new world, and to do it fast. We owe this to ourselves and to our children.

Related video: The Greatest Science Challenge in the History of Civilization

Then, 3000 years ago

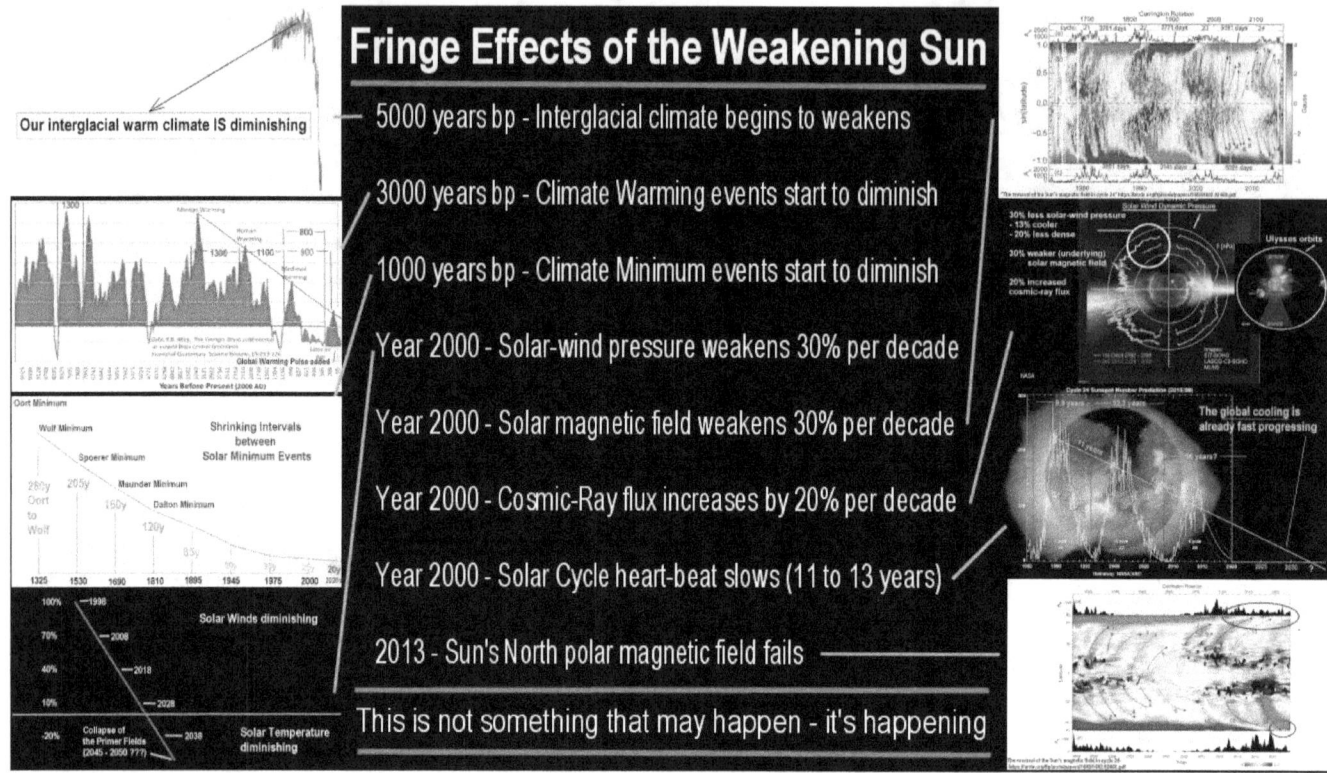

Then, 3000 years ago, the Climate Warming pulses and their intervals, started to diminish likewise.

After that 1000 years ago, the Climate Minimum events became weaker likewise, and their intervals shorter.

More recently, after the years 2000, or slightly before, the solar-wind pressure began to weaken at the amazing fast rate of 30% per decade, and so did the Sun's underlying magnetic field, as measured by the Ulysses spacecraft. In the same timeframe, till the end of the Ulysses Mission in 2008, a Cosmic-Ray flux increase of 20% was measured.

Also, in the same timeframe, the Solar Cycle heart-beat began to slow down, from 11 years per cycle to 13 years per cycle. And more recently, in 2013, the Sun's North polar magnetic field failed to materialize in the process of polarity reversal.

These are big fringe effects that are happening. They are not fantasies. They are measured events that have occurred. And we see more and more of them.

> The numerous fringe effects in solar dynamics indicate the solar system is rapidly diminishing towards a phase-shift point at which the glaciation starts anew on the Earth. Most of the fringe effects in solar dynamics affect us directly, except for the increase in solar cosmic-ray flux that results from the weakening Sun. The cosmic-ray flux controls our climate. It affects cloudiness massively, and also precipitation. It creates floods, and also drought. Increased cosmic-ray volume

creates droughts, by causing clouds to rain out faster, which reduces the water transport distance. It massive rainout occurs over land, floods result. Nearly all crop losses are caused by cosmic-ray showers and their consequences. That's what we need to pay attention to. The rest of the solar fringe effects are but alarm bells ringing that should get us to pay attention.

See: Floods and Ice Age

Measured in Carbon-14 isotope ratios

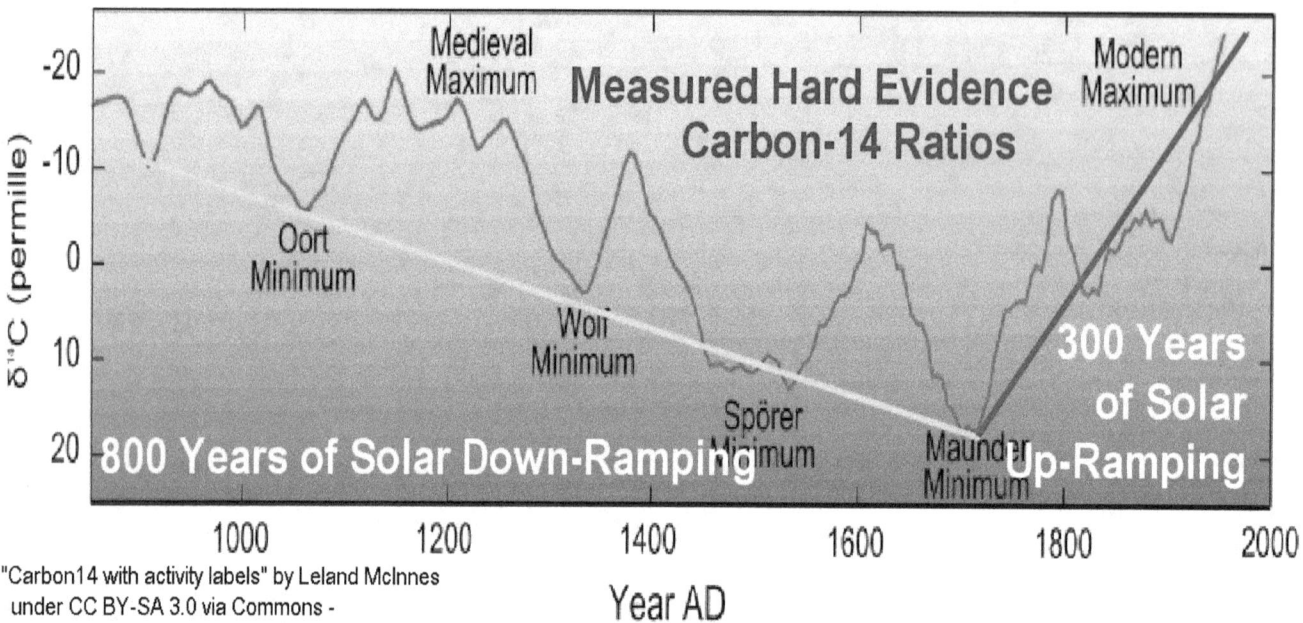

We have measured the solar activity diminishing for 800 years towards the 1700s, measured in Carbon-14 isotope ratios. And for the timeframe thereafter we have measured an up-ramping in solar activity for almost 300 years, with which the Sun gave us the famous years of global warming as a solar event.

> The time has come that we pay attention to solar dynamics rather than to politics that have become increasingly a circus of lies but on lies. We need to pay attention to what is real, because what is real affects the global food supply. One day cosmic reality will affect our politics. Till then, we have work to do in lifting ourselves up as human beings to redirect our perceptions that in turn become our politics and drive economic development.
>
> See: Cosmic Climate Change - Part 1 The Cosmic Forcing of society

The up-ramping of the Sun

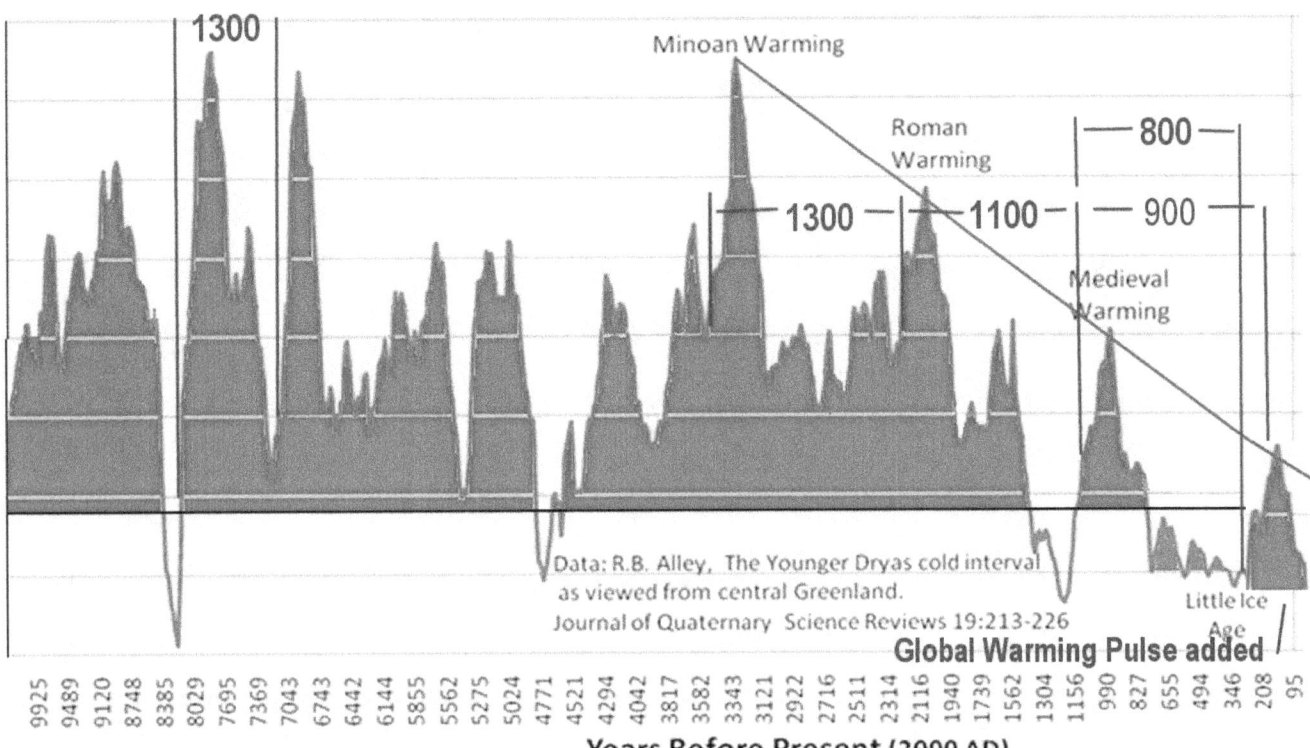

We also see evidence that the up-ramping of the Sun from 1715 onward, was caused by the cyclical occurrence of the big historic warming spikes, which have been getting progressively smaller for 3000 years already, and as I said, with shorter intervals between them.

The process is extensively explored in the video:
Grand Solar Minimum becomes the Ice Age (Part 2) - Uncertainty

Over the hump of the global warming uplift

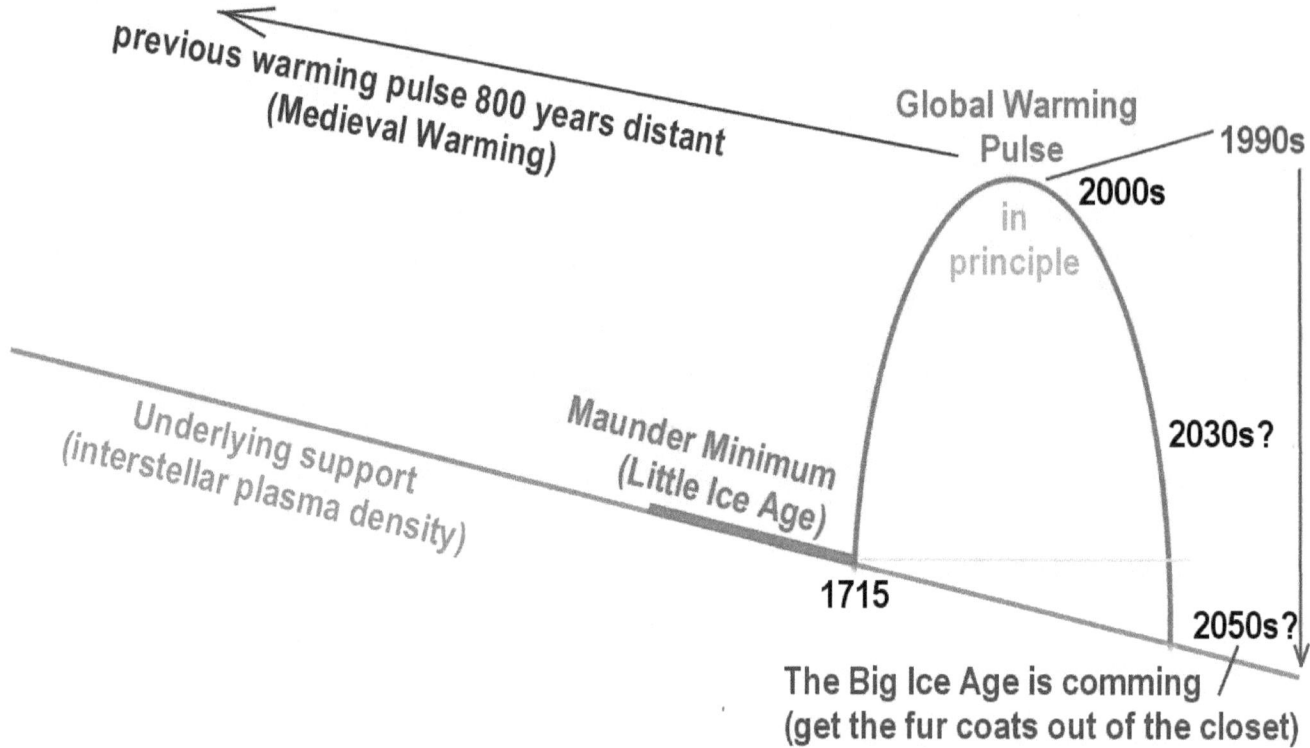

It became measurable, too, in the 1990s, that we are over the hump of the global warming uplift in solar activity, that got us out of the Little Ice Age. We are now on the down slope of the pulse that rescued us. We are nearly half-way down the slope, and we have measured the solar activity dropping off like a falling stone.

> This is the shape of the solar uplift that saved our existence with the last sequential global warming pulse. The universe gave us 300 years to prepare us for the big Ice Age. But nothing was built. The time we have been given is almost over. Shouldn't we at least start? To do nothing on this front, dooms humanity more certainly than a nuclear might, which we wouldn't likely survive either.
>
> More details of how this pulse was created is presented in the video:
> Grand Solar Minimum becomes the Ice Age (Part 2) - Uncertainty

Half-way along to the zero point

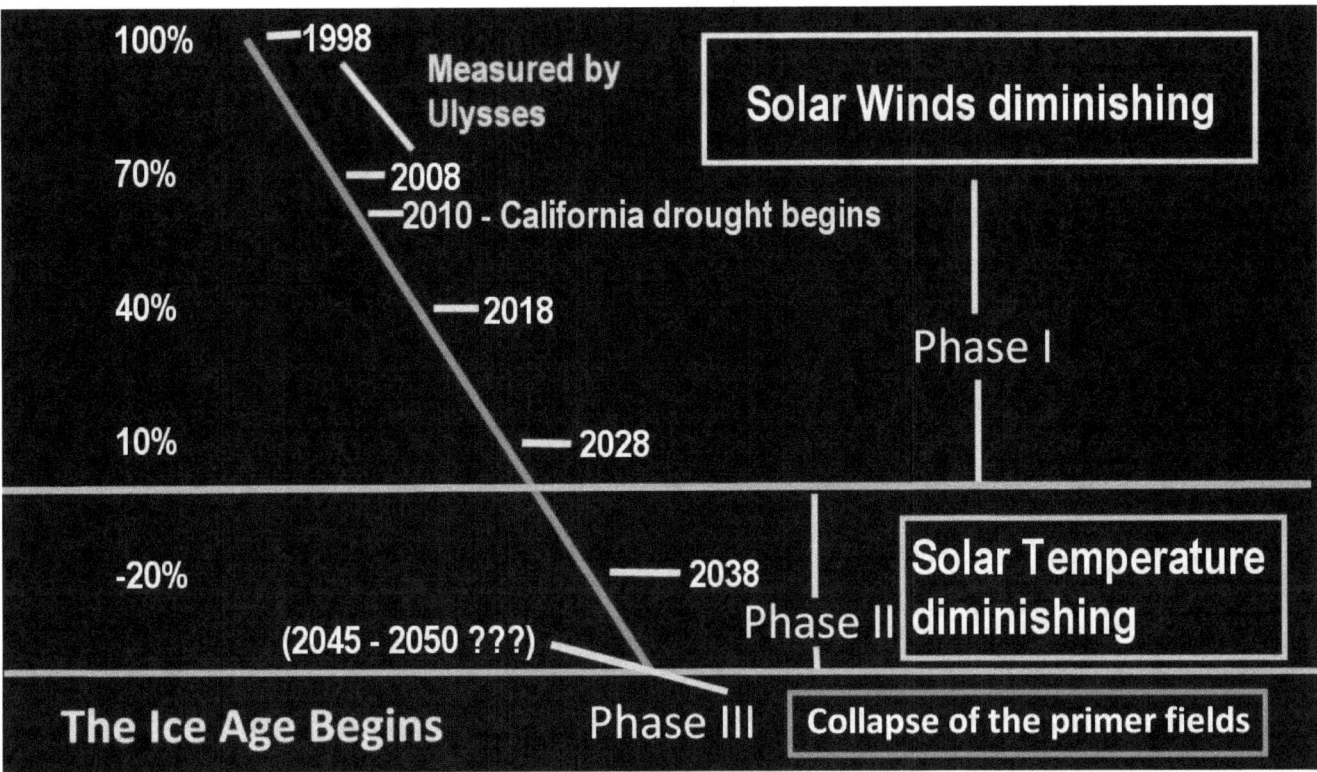

With the solar-wind pressure diminishing at 30% per decade from 1998 on, we find us already half-way along the path to the zero point, with the fringe effects increasing accordingly.

As I said before, past the zero point in solar-wind pressure, from the 2030s onward, any further weakening of the solar system will be expressed in the Sun's surface temperature diminishing towards the inevitable phase-shift at which the magnetic fields collapse that focus plasma onto the Sun.

> While the fringe effects mount up before we get to the Ice Age transition, the Ice Age itself is a digital-type ON/OFF event. When an Ice Age begins, the world becomes an entirely different place. Our task is to catch up with this potential and built us a new world that is fit for the new stage.
>
> See: Ice Age is Digital - Awake

The Sun falls back to its hibernation mode

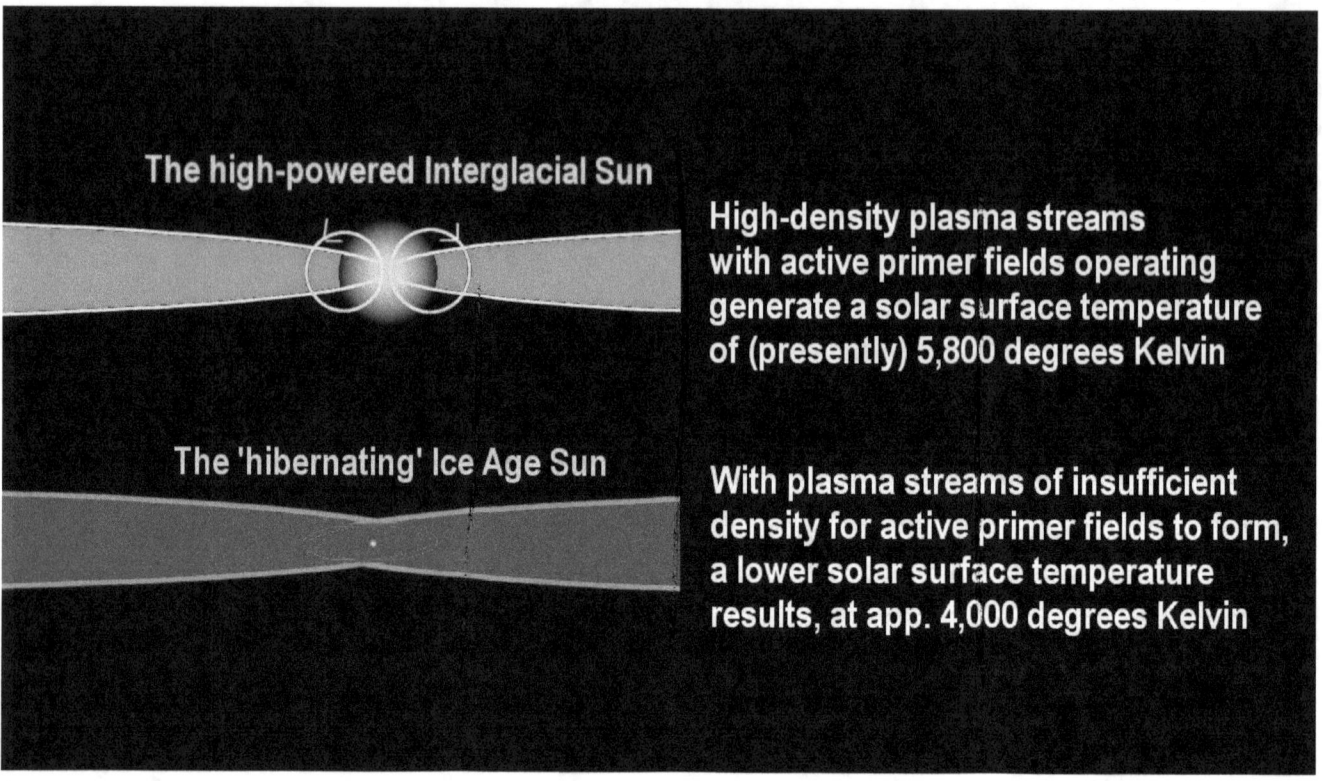

When this happens, the Sun falls back to its hibernation mode, a kind of default mode, a low-activity mode that produces 70% less radiated energy.

> Ice Ages result when the galaxy is weak, and consequently the interstellar plasma streams become too weak to maintain the Sun in its current high-power mode, which covers only 15% of the Ice Age cycle. When this mode ends, a default mode begins in which the Sun is 70% less active - a type of hibernation mode. Are we willing to give our children a chance to live under a hibernating Sun?
>
> See: National Security: Children at Risk

The phase shift is inevitable

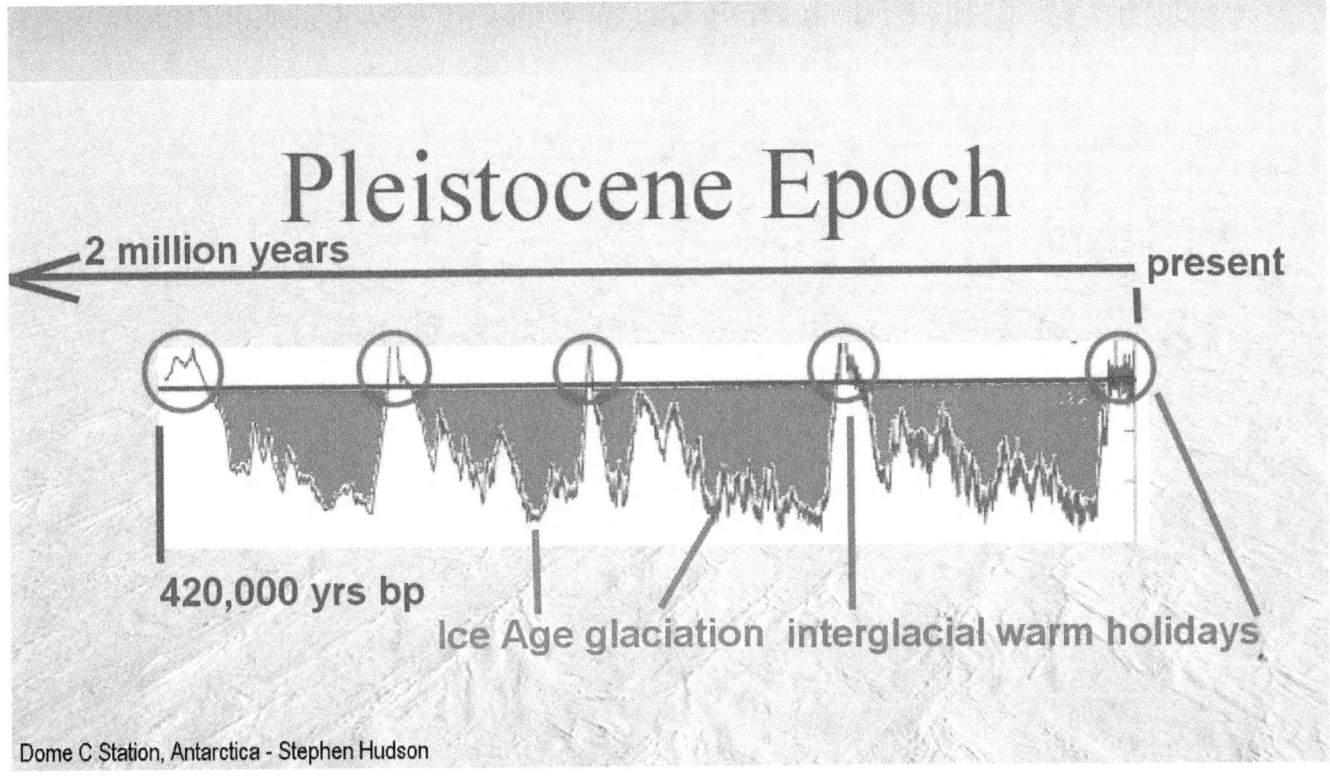

The phase shift is inevitable. It has happened before, by the cosmic principles involved, and will happen again. Only the exact year is somewhat uncertain.

When we speak of an Ice Age, we speak of a different world where all the meters are different. Nothing remains the same when we enter the glaciation world, nor will there be gradual blending between these two extreme opposites. The transition is digital. It is an on/off transition. The yardsticks with which we measure the interglacial variations, such as floods and droughts, no longer have any meaning in the glacial world. While the current interglacial climate is fading out, all that we experience is still a phase of the interglacial climate when the Sun operates in its high-powered mode, even the worst of the fringe effects.

If we don't recognize the digital nature of the Ice Age transition, we fail to recognize the nature of what we are getting into and fail to prepare for it. In like manner do we fail to make the radical shifts in thinking and in relationships with one another. A complete uplift of our humanity is required, far beyond just shifting from big wars to little wars. We need a phase shift in regarding our universal humanity to match the enormous scope of the climate change before us. If we aim for anything that falls short of a total renaissance environment, then we have already lost, and promote the murdering of our children by indifference and worse.

A test of our sincerity is found in our commitment to provide the 6,000 new cities for one another for free, as a starting gate to a richer and brighter and more productive world. If we fail this test, we won't make the grade and the future is lost.

See the video: [Life Blocked: Digital Ice Age Denied](#)

Timing is determined by the threshold level

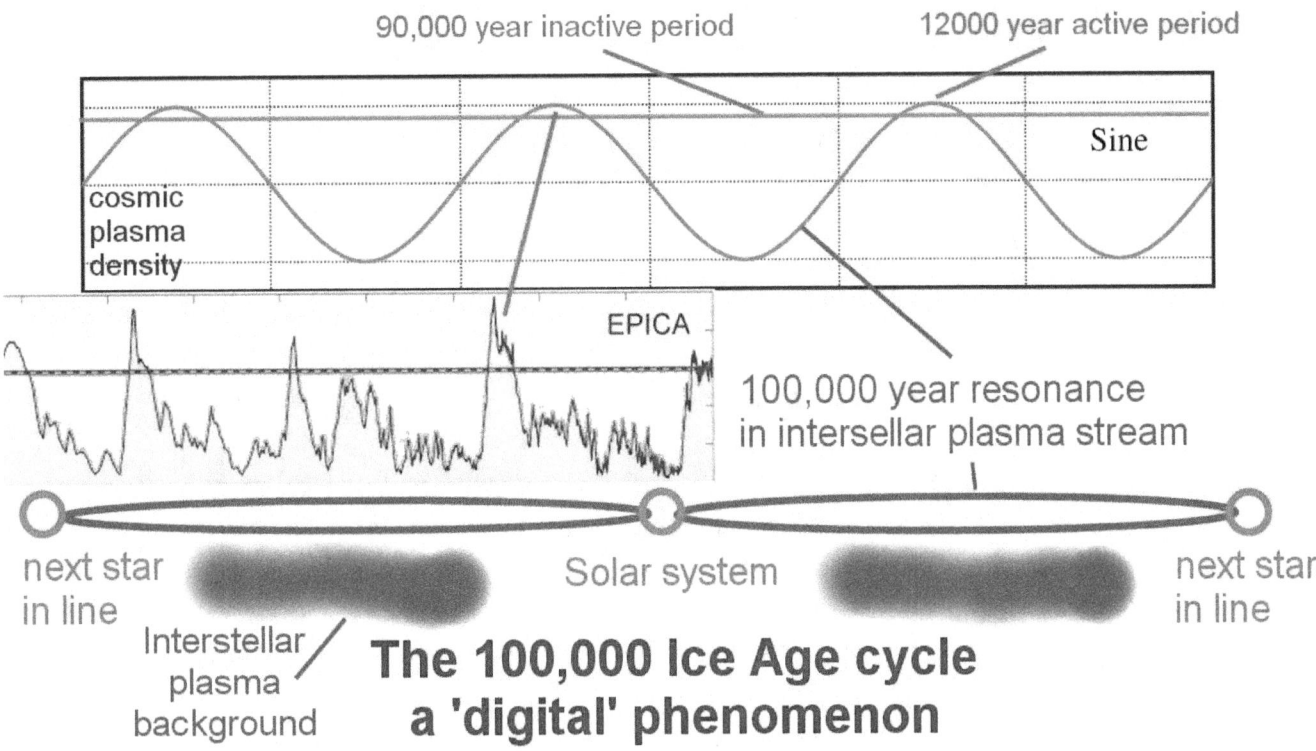

The timing is determined by the threshold level at which the magnetic fields collapse, which presently focus high-density plasma onto the Sun. The threshold line determines the turn-on timing, and the turn-off timing, for the high-power period of the Sun, its high-activity period, which generates the interglacial climate on Earth.

> Few people recognize that our current climate, the interglacial climate, is a climate anomaly. For more than 85% of the last half-million years that we have ice core records from, the Earth has been an virtual Ice Planet, with life being sparse on it. We call the current interglacial, the Holocene Epoch. We have lived in a wonderfully warm holiday for the last 12,500 years. We had this holiday enabled by the interstellar plasma streams reaching their peak density in their resonance cycle. We are about to drop below the threshold value needed for the Sun to remain in its high-power mode. We are about to drop out of this mode as the peak of the resonance is over.
>
> This means that we are about to drop out of the warm climate environment, and back to what has been normal for most of the last half a million years. Except we don't know what 'normal' is, in humanist terms.
>
> I suspect that the 'normal' in humanist terms, is found in all the qualities of our humanity that are qualities that are enduring, such as universal love, cooperation,

generosity, mutual support, and the like, which are qualities that meet all human need. Against this background of the enduring, the imperial environment and its attributes of slavery, monetarism, terror, fascism, stealing, war, hate, torture, and so on, are falsely cultivated attributes that are not of the scene of what is normal and enduring of the human dimension, or else humanity would have ceased to exist long ago. This difficulty of stepping away from this entire group of cultivated anomalies needs to be corrected, and this has not yet been achieved, though some progress has been made.

Humanity has rarely stepped away from the universally destructive background, and in the few cases when it did, great periods of renaissance did unfold, but only for them to become overpowered again.

It can be said with relative certainty, that if this train to hell isn't stopped decisively and intelligently, and a permanent humanist renaissance is established, society can kiss its future good bye, whereby history ends.

This is a part of the challenge that the Ice Age Challenge includes. It bits us to rise up to that challenge, or else the world-ship is doomed.

See: Arresting the Infinite Crime

The rapid weakening that we see

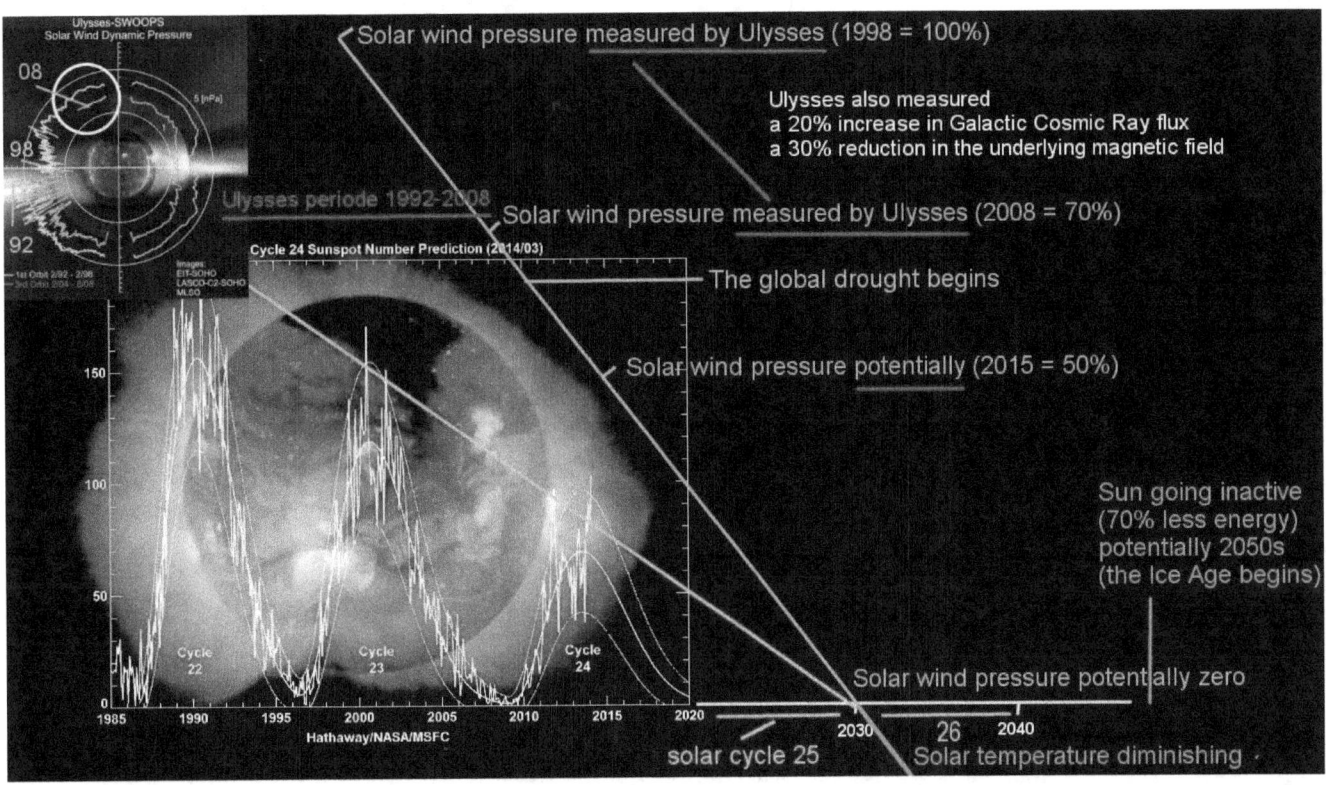

The rapid weakening that we see now evident in so many ways in the solar system, as diminishing solar wind, diminishing sunspot numbers, slowing solar cycles, and fading magnetic fields, and so on, indicates that we are quite close to the turn-off point.

A viewer, (Marc C), commented that the same pattern, though with a steeper collapse, is also observed in Radio Solar Flux measurements. He notes that a normal solar maximum produces a flux factor exceeding 200, but that the last maximum (assuming 2014) was measured at only 112 in volume, a mere half of what it used to be. He also notes that the last solar minimum (assuming 2008) produced a factor of 59.

- Today's flux density measurement (4/17 - 2018) produced a factor of 69.8, while we are still 6 years away from the solar minimum for the current solar cycle, cycle 24. (see: http://spaceweather.gc.ca -select solar flux)
- The sunspot numbers have collapsed in a similar manner, from a maximum of 212 for cycle 22, to a mere 81.8 for cycle 24 (Apr. 2014) - a 62% collapse.

The viewer commented: "Because the Sun is operating at half power already you can expect the sun dying when the full solar minimum hits in 2024. The Sun/earth will not recover from this. From this year forward, 2018, every year will become colder. Leave Europe within 3 years, before all borders close."

Of course "the sun is dying" is just a phrase to wake people up. The Sun is merely a type of energy converter. What flows in determines what flows out. However, the interglacial climate is in a process of dying. We know from Ice Core records what the climate will become when the glaciation starts, but we have no historic yardstick to measure the transition period. While we have still 30 years left to go till the phase shift, climate related crop losses are already increasing amazingly fast in some parts of the world. Seeding is delayed by cold and snow, and in some places by too much rain. When the growing season is shortened by several weeks that way, and the ripening season is shortened by early frosts, both of which are already experienced, crop failures will dramatically increase and food prices become sky-high.

While the crisis can be avoided with building large-scale indoor agriculture, nothing of this sort is being considered, much less a real solution, such as creating large-scale tropical agriculture afloat on the equatorial sea. The official answer is: Let the people die! Depopulation is good.

China's answer is, to aid the economic development of tropical Africa. In the mean time, the Sun is diminishing and the world drops ever deeper into the pit of war, destruction, and terror. Nevertheless, I like to think that as human beings, when we raise ourselves up to become that, will see ourselves 'forced' by the changing cosmic climate to accept the technological solutions that are opened to us for secure human living under the worst conditions, whether it the be floods, earthquakes, drought, cold. Indoor agricultures afloat on the tropical sea is save from all of these.

That's the gist of my video:
Cosmic Climate Change - Part 2 - From Big, to Gigantic, to Joy

These things are happening now

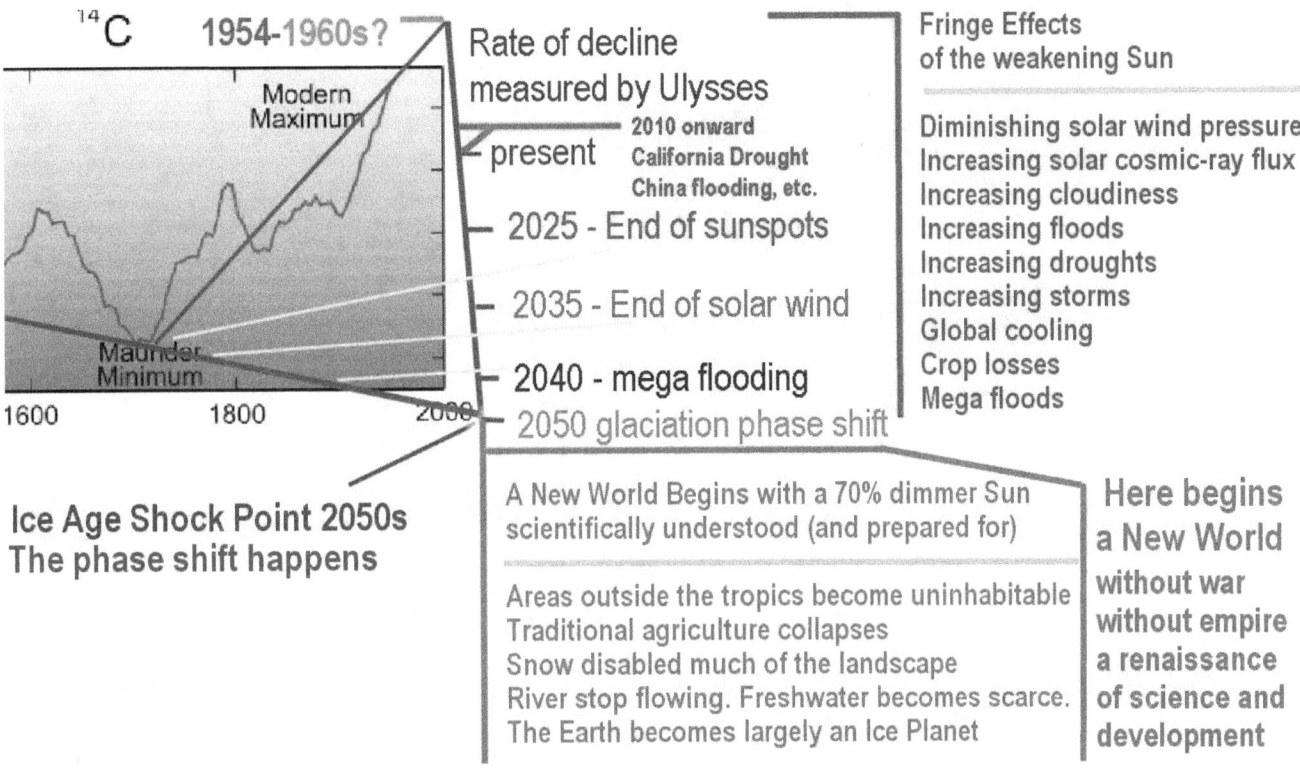

These things are happening now, not a thousand years from now. Global cooling, floods, storms, droughts, are all increasing at a rapid rate and will continue to increase for 30 more years with evermore crop failures along the way, until the phase shift happens and the Earth becomes uninhabitable without massive technological infrastructures, such as floating indoor agriculture with artificial environments, that have the power to supply the human need for 7 billion people, effectively, on an otherwise uninhabitable planet.

> The cosmic solar system is changing amazingly fast. Weed need to move with it. This means that in order to meet the Ice Age Challenge we need to develop out civilization now, wile we still can, in order to meet the larger goal.
>
> This means that the building of infrastructures to secure human living on this planet, such as the building of 6000 new cities with protected indoor agriculture that we will ultimately need, needs to start now. But why is this not even considered? We have known since 2008, measured by the Ulysses spacecraft, that we would be in a food crisis in due course, but we have done nothing. The bottleneck on this scene is not the physical challenge, nor scientific knowledge, but society's indifference to what is already plainly known.
>
> This is the gist of my video: Ice Age is Digital - Awake

When the Earth becomes uninhabitable

World map with the intertropical zone highlighted

By KVDP - Own work, CC BY-SA 3.0, https://commons.wikimedia.org/w/index.php?curid=27385077

When the Earth becomes uninhabitable in the 2050s, and the zone for the New World to be built shrinks to the band of the tropics - when the territories of Canada, Europe, Russia, the USA, China, and parts of India, become unsuitable for human living, and the remaining areas becomes disabled by drought, - the pages of history take on a new color, new pages become written as the physical foundation for our existence shifts almost entirely from living off the land, to living by technologically created resources, with much of those located on the sea.

> We are not at the point yet when the Earth becomes uninhabitable by today's standard. This means that our standard is too low, too small, too impotent. We embrace the option of failure. Society throws up its hands and capitulates. That's not a fit attitude for winning the war against indifference for building a New World. - The issue is survival with 7 billon people on a planet that becomes uninhabitable by today's standards.

> Failure is not and option:
> Dynamics of Wisdom in an Ice Age World

> In order to win on this front, we need to uplift the entire world-political system. If the system fails us, let's uplift it to a higher level where it serves our goal to have a future. Some small steps have already been taken along the way, which presently may seem like giant steps, though they are feeble in comparison with what we can

accomplish as human beings when we raise us out of the rut with cultural development. China and Russia are already, daringly, hacking into the decaying landscape of western imperialism and its war-focused degeneracy, to uplift it.

Cultural development sets up a new standard that isn't defined by space, but by humanity that creates its own space and environment, both in beauty and infinity.

This is the gist of my lengthy video: <u>Yes, the Russians did hack</u>

A crisis in history unfolds at the present stage

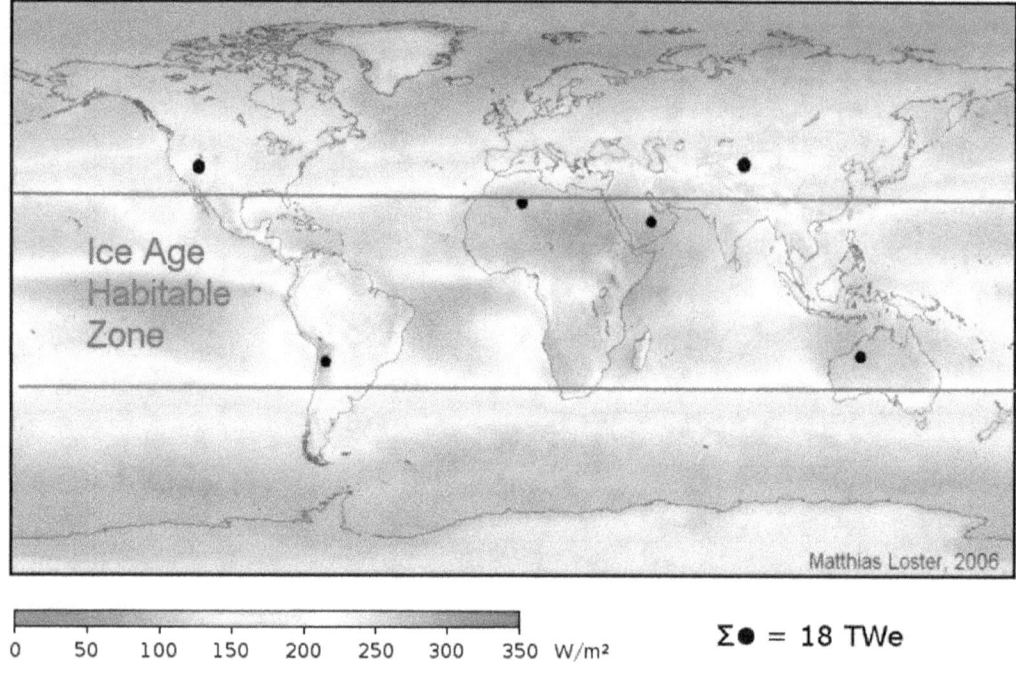

 A crisis in history unfolds at the present stage, because the artificial resources for continued human living need yet to be developed, which is totally ignored. And the infrastructures need to be fully operational when the phase shift happens. We are at an existential crisis, because the development of these existentially-critical resources is not even considered anywhere in the world, much less is it being implemented.

The giant is asleep

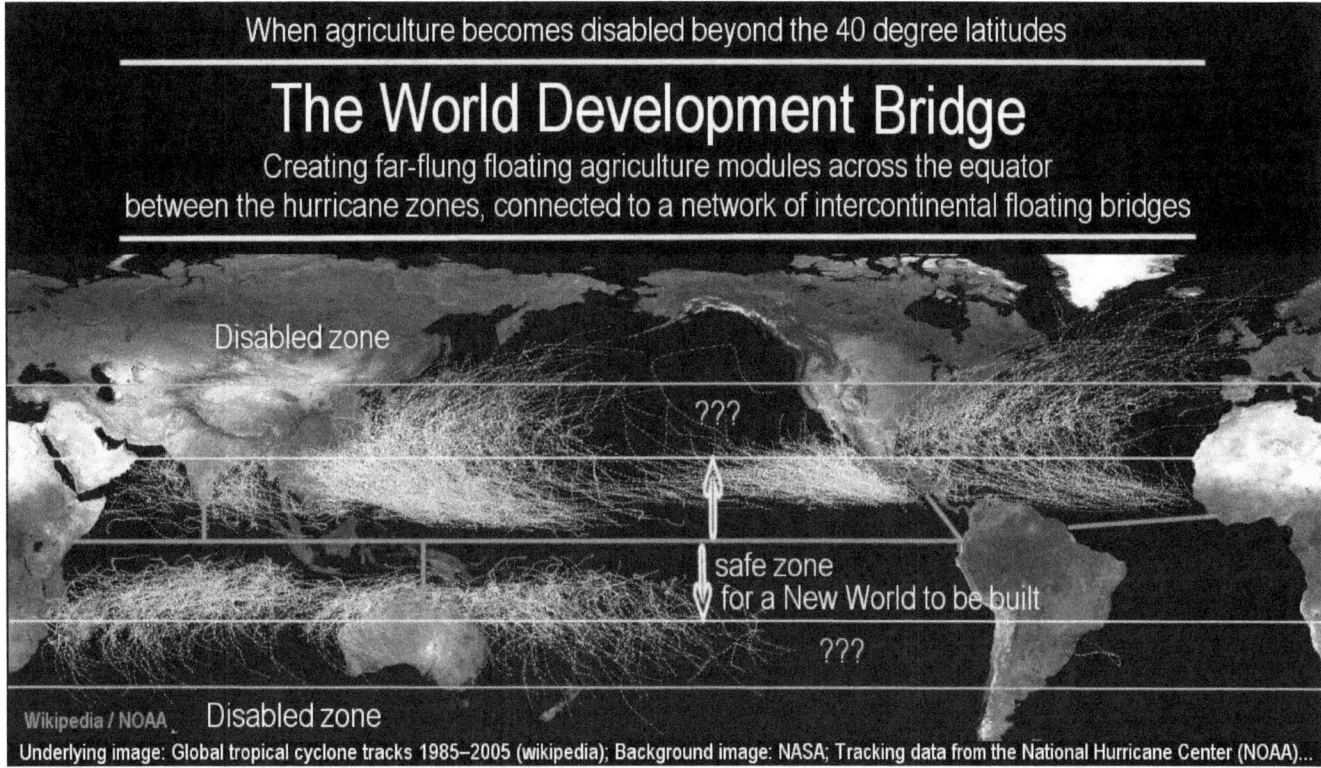

The building of the infrastructures for the required new resources should have already started. It needs to start now. It needs to be implemented faster than the interglacial system is diminishing. But nothing is being done. The creative and productive giant that humanity inherently is, that it has the potential to be, is asleep. Even its children are asleep.

A city downstream from a great dam

reservoir and Merowe Dam in Sudan.

Our world is comparable to a city located downstream from a great dam that is in the process of breaking up. Deep fissures appeared. The fissures are getting larger. Fountains of water are spewing from the bigger ones. A concerned person stands in the hustings, suggesting to everyone that they band together and build a new city on higher ground where the breaking of the dam would not affect them.

In our modern world, many would come to hear the warning, but no one would be interested enough to put the spate into the ground to start building. In the story, some said that they didn't have the time, they had a business to run, a wedding to prepare, a farm to tend to, and a bakery, a butchery, and an automobile repair shop to operate. Most refused to even acknowledge that the dam is breaking up, and those who are able to see the obvious, said that the breaking of the dam isn't their concern, and refused to get involved.

"But you are involved" argued the man back from the hustings. "No one is not involved. One is either involved in saving the people of the city by relocating it, or one is involved in assuring the death of everyone in the city with the breaking of the dam, which is the result of doing nothing. None-involvement is simply not possible in this case."

The Ice Age is close to be breaking

The outcome of the story remains yet to unfold. It is our story. It is the story of the current world. The Ice Age is close to be breaking upon us. The evidence is already being felt with evermore severe fringe effects in the form of floods, droughts, tornadoes, hurricanes, freezing cold, even earthquakes.

Some call the effects, the effects of global warming. Others say that the effects are temporary effects of a Mini Ice Age or a Grand Solar Minimum that will come and pass by, as it did in the 1600s.

No one wants the hear the word "Ice Age" being spoken, seriously. Society loves its dreaming, as it is told that the real Ice Age won't be happening yet for thousands of years. Consequently nothing gets started to stay on track with the universe, nothing gets done that is vital for our continued existence. There is not enough love left in society for its humanity to bridge the many gaps of denials.

➢ Humanity is riding the 'Merry Go Round'

Humanity is riding the 'Merry Go Round' of trivial pursuits, while its future slips out of its reach.

My numerous science-video productions

Against this background, of the now fast unfolding existential crisis in civilization that humanity has become indifferent to, I have made a number of extensive efforts to explore the science of the Ice Age dynamics, and have presented the results in the form of my numerous science-video productions.

The metaphor of the carousel

The video presented here, is designed as an overview, but one that is extensively indexed in the transcript pages to the previously produced videos that explore the numerous aspects of the Ice Age dynamics.

I have chosen the metaphor of the carousel to hint that in the Ice Age explorations new vistas come into view, as the carousel rotates.

The Ice Age Challenge a bugle call to us

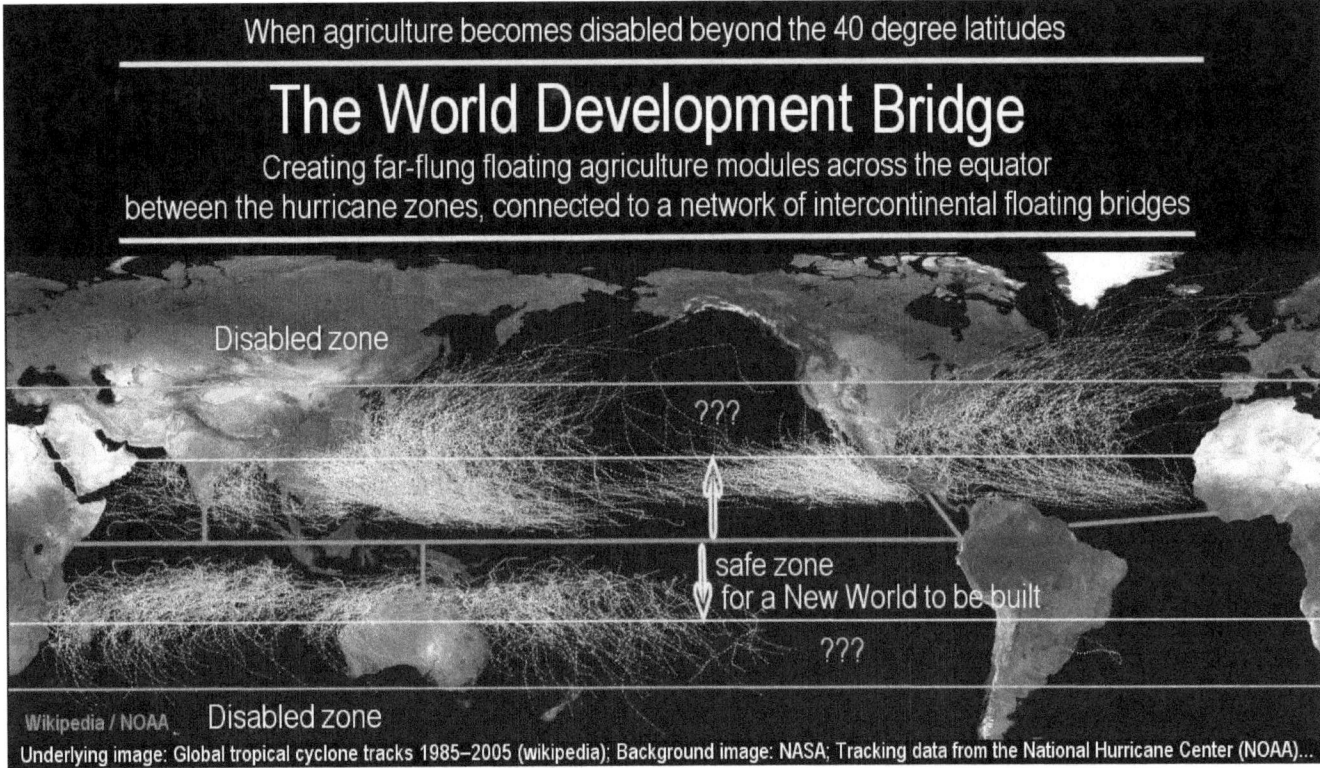

I also need to add that I do not present my videos as Ice Age scare stories. I aim to uplift the Ice Age Challenge to the level of a bugle call to us, by the universe, to rouse us to build us a new world with technological infrastructures that the glaciation consequences cannot affect. This, we can most certainly do, and thereby end up with a richer, more-secure, and with a more beautiful civilization than we presently have, with greater happiness and satisfaction in living, and one that assures us a future.

What do we know about the Ice Age dynamics?

What do we know about the Ice Age dynamics?

We know far too little.
But what we know is amazing.
It is powerfully relevant.
It is hugely significant.

What do we know about the Ice Age dynamics?

We know far too little. But what we know is amazing. It is powerfully relevant. It is hugely significant.

Most people are fast asleep

Most people are fast asleep today in the face of the greatest danger to their existence, just as the people in the 1600s were unaware of the scope of the cosmic dynamics when the Little Ice Age erupted that had affected their agriculture and caused wide-spread starvation.

No one had any idea then, of the great danger that the whole of humanity was in at the time, and how close the Earth had been to the phase shift to the full Ice Age that no one was prepared for, and no one would have survived.

In the ice core record

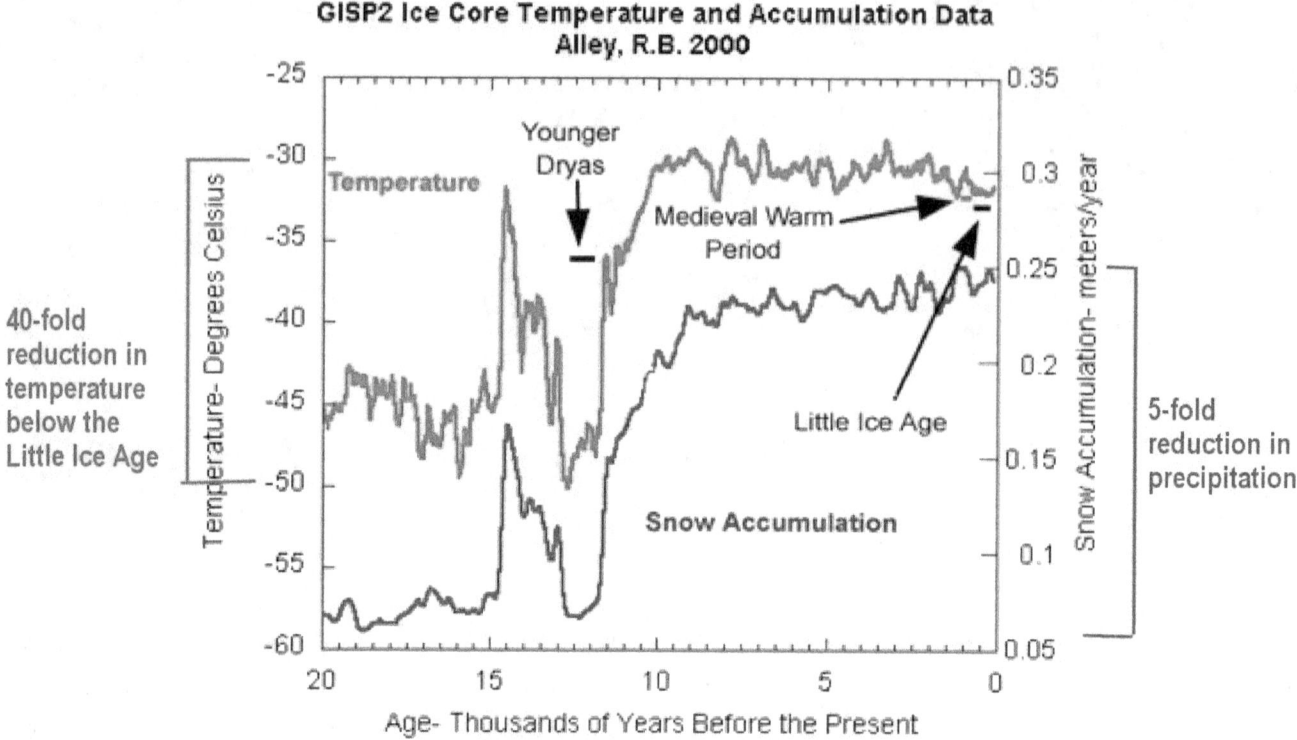

It wasn't until into the 2000s that ice core data from Greenland revealed that the Earth had been essentially uninhabitable during the last glaciation period.

In the ice core record, the Little Ice Age shows up as a half a degree of global cooling below the 2000 average, while the full Ice Age is recorded as having been 20 degrees colder. This adds up to a 40-times deeper cooling in comparison. And more than this, the precipitation is recorded in the ice core samples as having been 80% less in glacial times, than it is today. No one could have imagined this in the 1600s, much less had been prepared for it.

Immense cold under glaciation conditions

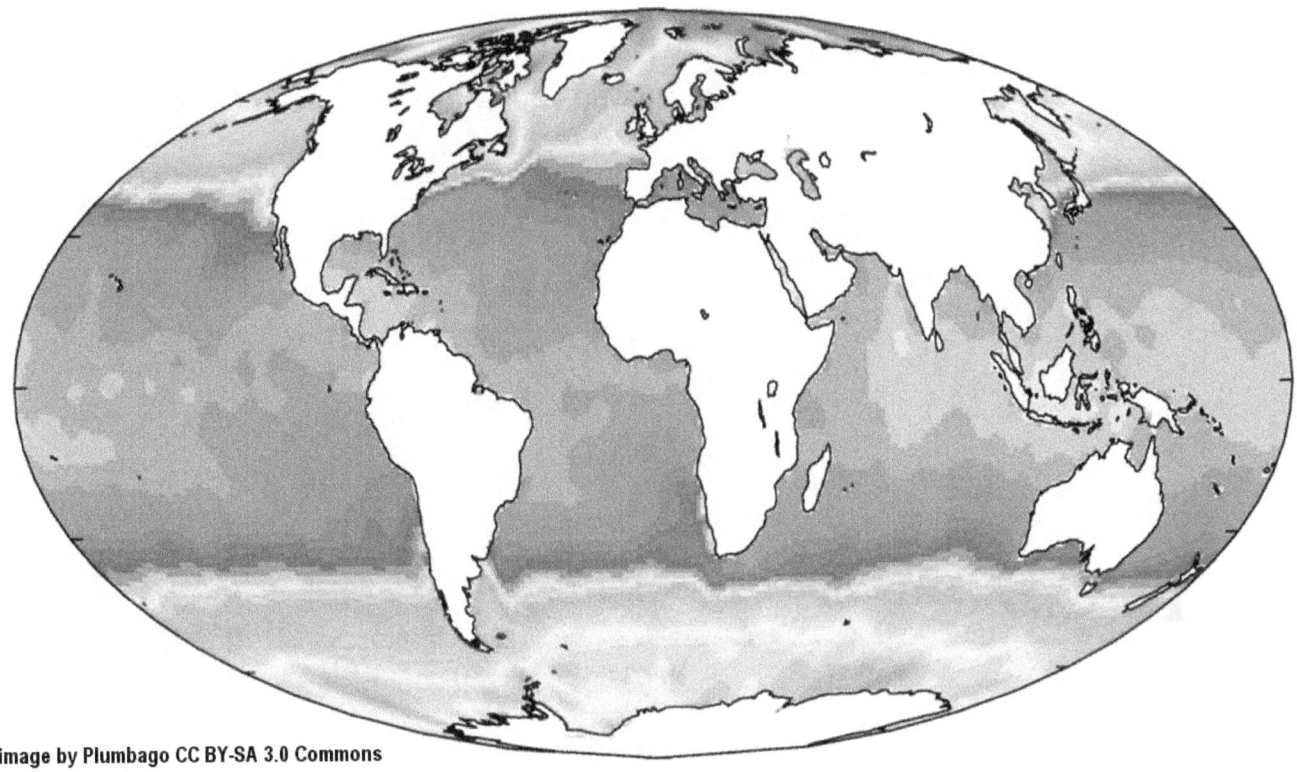

image by Plumbago CC BY-SA 3.0 Commons

All the lands outside the tropics, outside the pink zone, are lands that become reclaimed by the resulting immense cold under glaciation conditions. Agriculture is impossible there, and without it, human living is impossible even before the snow piles up into deep sheets.

And the lands within the tropics that remain ice free under a colder and dimmer Sun, become total deserts, with precipitation diminishing by more than 80%. Without rain, the entire freshwater infrastructure stops. Agriculture as we know it, even in the tropics stops, for the lack of freshwater.

The up-ramping of the Sun in 1715

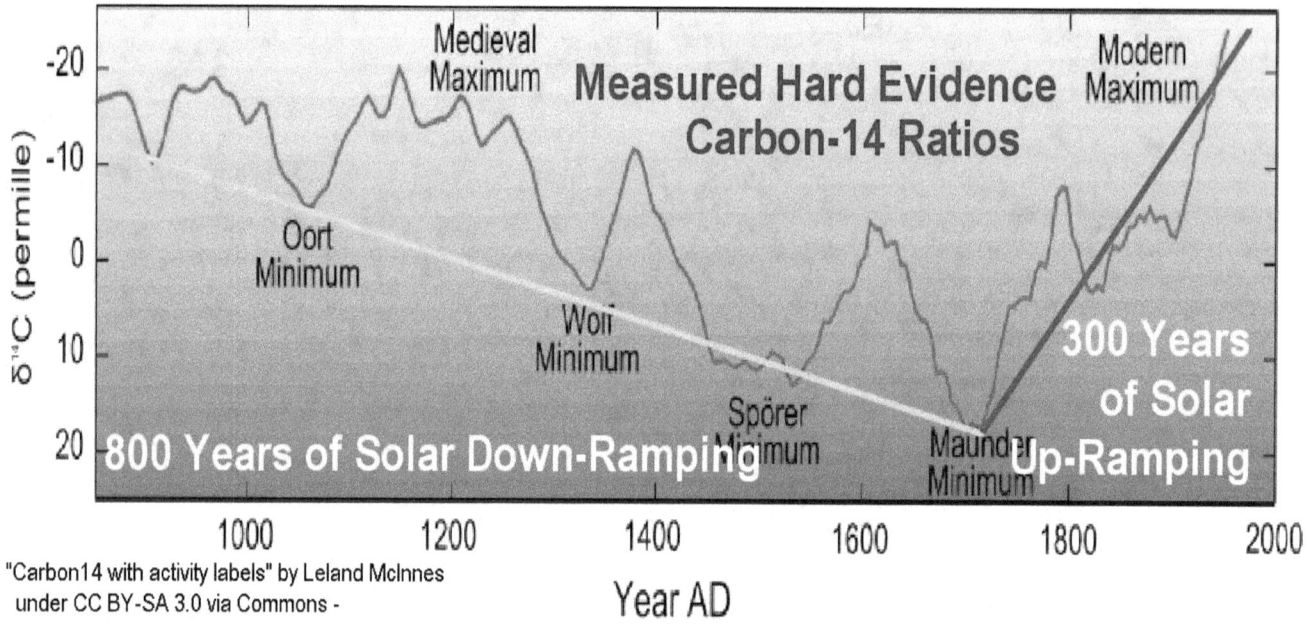

"Carbon14 with activity labels" by Leland McInnes under CC BY-SA 3.0 via Commons -

As I said before, fortunately for us all, the Sun was ramped up again in 1715, in a big way, whereby the underlying weakening to ever-lower levels of solar activity, and the consequent full Ice Age phase shift, was prevented. The up-ramping of the Sun in 1715 gave us the nearly 300 years of global warming that politics, and even some scientists, blame humanity for, and wreck its economies to prevent it, even while this life-saving global warming was clearly caused by increased solar activity. Carbon-14 ratios stand as measurable proxies for historic solar activity, which prove that the global warming after the Little Ice Age had been caused by the Sun.

Regardless of the politics involved, the wonderful global warming that had lasted all the way into the 1990s had been measurably a solar event that humanity doesn't have the power to affect in any way. The cause for the solar event is built into the dynamics of the solar system as a whole.

Without it, none of us would exist

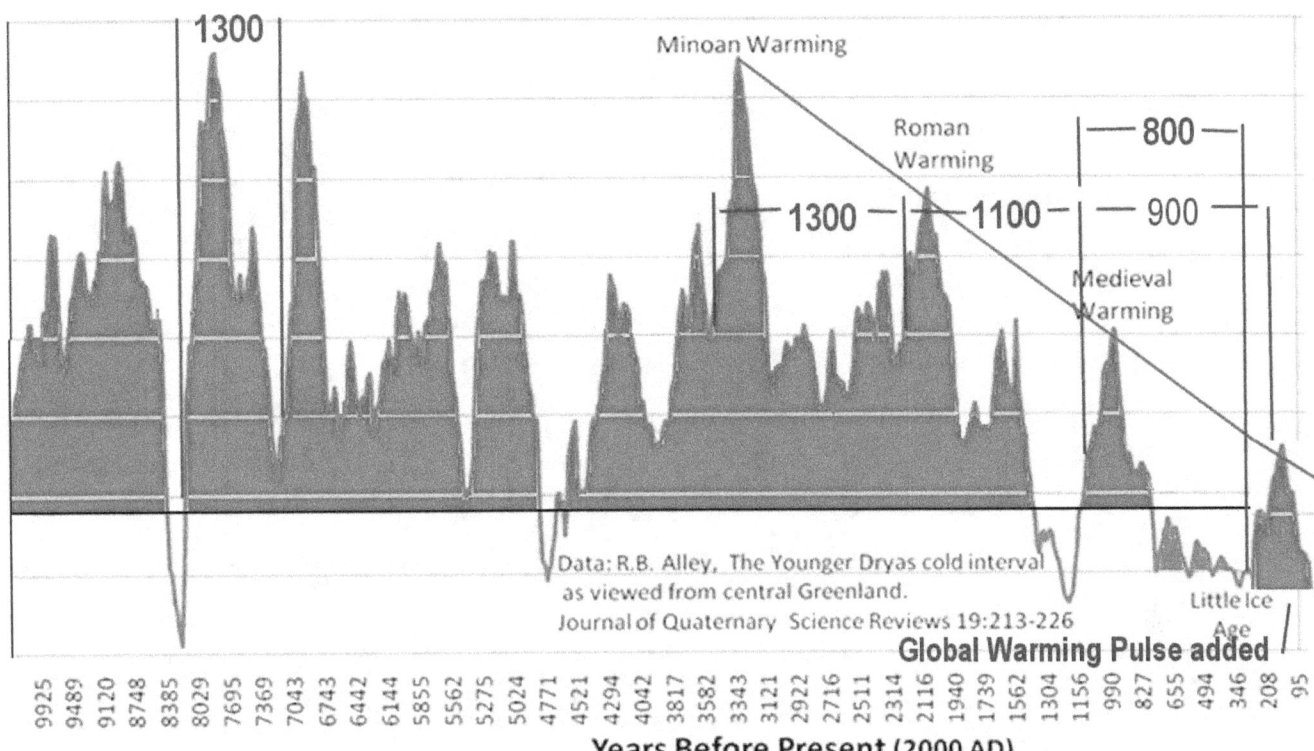

As I also said before, the cause for the up-lifting of the Sun, that had stopped the little Ice Age and gave us global warming instead, was the dynamic continuation of the three big historic climate warming pulses. As these pulses have been getting progressively smaller in amplitude, the intervals between them have been getting progressively shorter. The intervals had diminished from 1300 years, to 1100 years, and then to 800 years. It was the last event in the series that saved the world in 1715. Without it, none of us would likely exist. No one can live without food. When glaciation begins, agriculture as we know it ceases to be possible. This means that food production stops globally under glaciation conditions.

The world population of the last Ice Age

The world population that came out of the last Ice Age, the few who survived in isolated pockets, probably lived of fish. This is possible for minuscule populations, and is evidently precarious. Ice Ages in the past have been so harsh that humanity nearly became extinct during the second-last Ice Age.

Now, we are 7 billion people

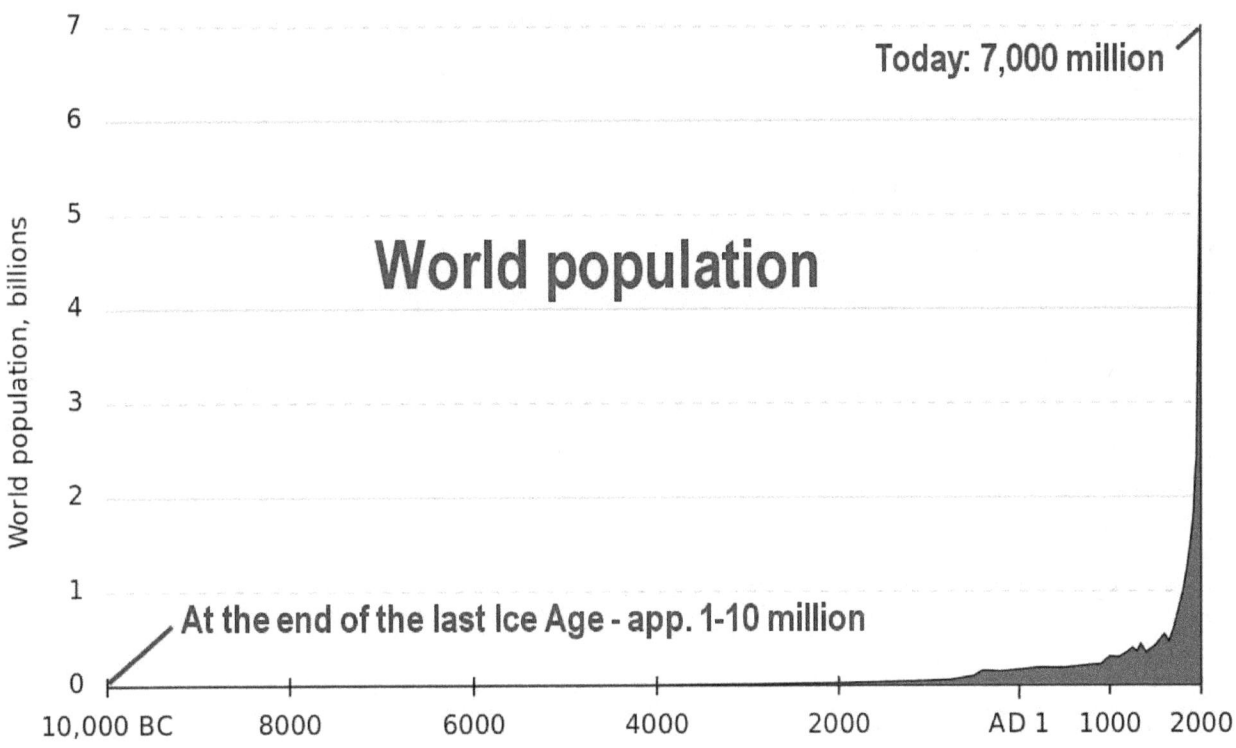

Evidently, humanity didn't fare much better during the last Ice Age that had yielded a mere 1 to 10 million people worldwide. Now, we are 7 billion people. We have the same harsh glaciation knocking at our door, potentially in the 2050s, that humanity had barely survived in extremely small numbers. Obviously, we want to do much better than that.

The entire development of civilization

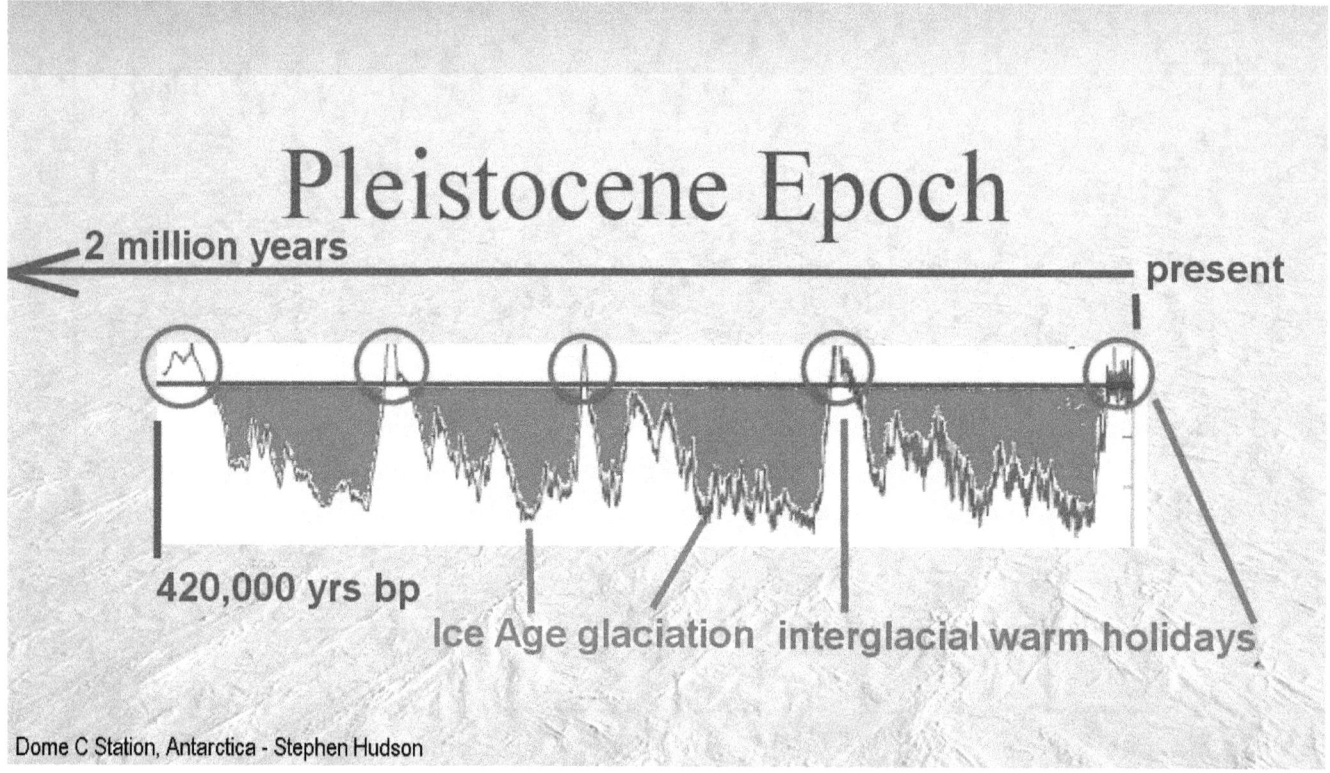

All that we have become - the entire development of civilization - occurred in the current interglacial period, the current period between the glacial periods. Never before in history has humanity had a 7,000 million world population, which it cannot allow to dwindle back to the 1 million level.

➤ **We need to build us a new world in the tropics**

We need to build us a new world in the tropics.

We have no choice. But will we do it?

We need to build us a new world in the tropics.

We have no choice. But will we do it?

We need to build 6,000 new cities

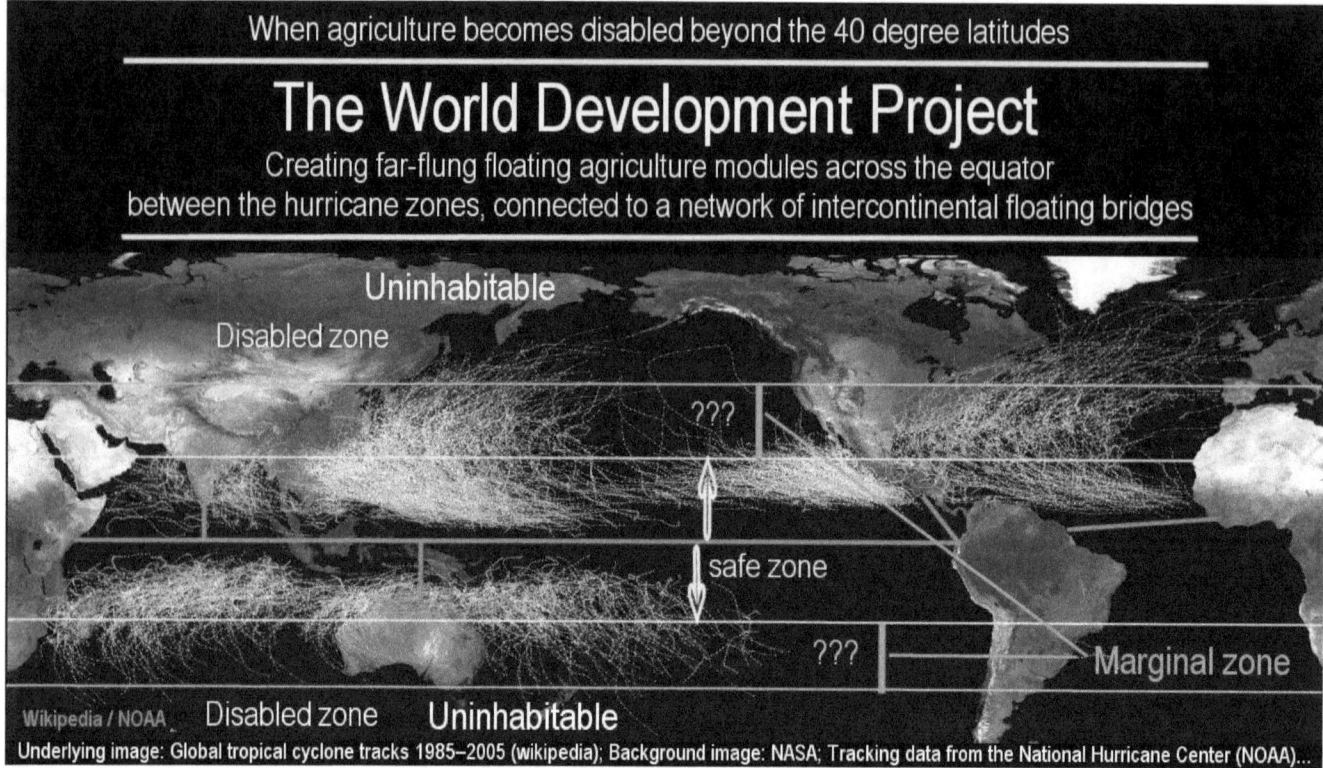

We have the capability and the resources to build us a new world in the tropics, that the Ice Age cannot touch. We can accomplish this with the most modern types of technological infrastructures, with which we can support us with a richer and more powerful civilization. But nothing of this sort is even considered, much less is it being built.

For meeting the Ice Age Challenge, we need to build 6,000 new cities for a million people each, deep in the tropics, preferable along the equator, with new agriculture attached to the new cities, that is large enough to nourish 7 billion. people. And since suitable land is scarce in the tropics, the new infrastructures will need to be placed afloat across the equatorial seas. That's the challenge. It is not a small step to meet this challenge.

But why should we fail on this front? The prospects that we find on this front are tremendous. Building a new world across the sea, with new cities and new agriculture, is an open door to greater prosperity and happiness than we have today.

All this is imminently achievable. Why then would anyone opt for death by starvation by doing nothing to prevent it, when a richer world than yet imagined, lays within our grasp? Let's reach for it and live.

➢ A Wellsian Crisis Project: To disable science

The barrier that blocks the Ice Age Renaissance.

A Wellsian Crisis Project: To disable science

The Barrier that blocks the Ice Age Renaissance.

A Wellsian Crisis Project: To disable science.

To stop the march of science by all means

Herbert George Wells in 1943
the man who changed
the course of science

The barrier that presently blocks this path to a secure Ice Age World, is not a physical barrier, but is a political one.

The science that would enable society to understand the Ice Age dynamics, and its potential start-up in roughly 30 years, is at the present time politically blocked. It is blocked by an ideological project that is designed to keep a dying and corrupt imperial system alive that cannot maintain itself by its own resources, but relies on stealing. Its goal is to stop the march of science by all means possible, for the purpose of keeping humanity impotent.

scrp-awake/tafa024.htm

H. G. Wells wrote the script for the process

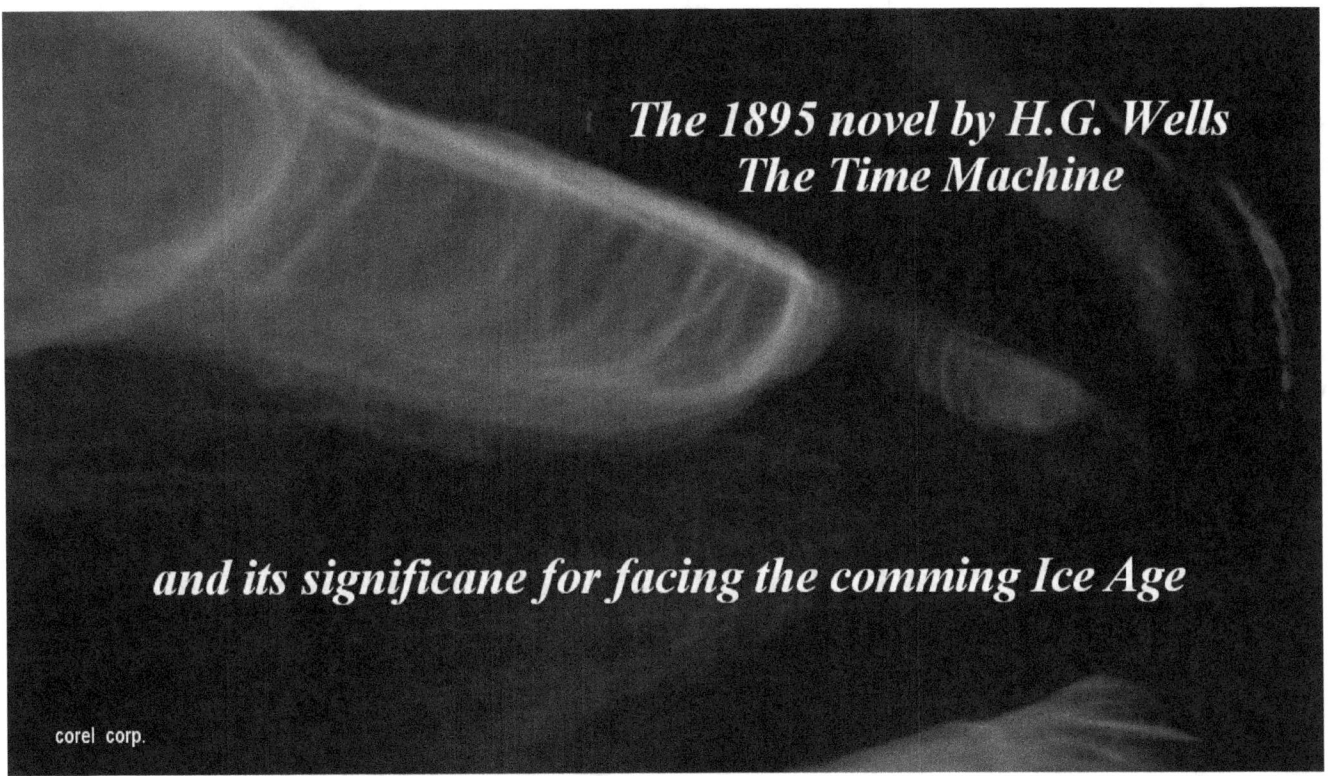

H. G. Wells wrote the script for the process, thinly veiled in his novel, the Time Machine. The story of the novel takes its hero, an inventor of a time machine, far into the future. There he encounters an elegant, but docile people, named the Eloi. The Eloi are the former noble elite living in an idyllic world with plenty of food, who have no need to produce anything. The traveller also discovers what keeps this world running. He discovers the Morloch, the equivalent of an industrial scientific society, the masters of machinery who operate the productive processes. The traveller discovers their secret. The science and technology people keep the Eloi as livestock. Wells' message to the Elite, was, not to allow science to gain the upper hand, or else these practical science people will eat you for breakfast.

scrp-awake/tafa020.htm

A debate as to how to deal with science

Herbert George Wells in 1943
the man who changed
the course of science

There was a debate on the subject in the 1920s, as to how to deal with science. Apparently it was decided that science can be effectively crippled with irrationality, and thereby be controlled.

The Hydrogen-Fusion Sun doctrine was invented

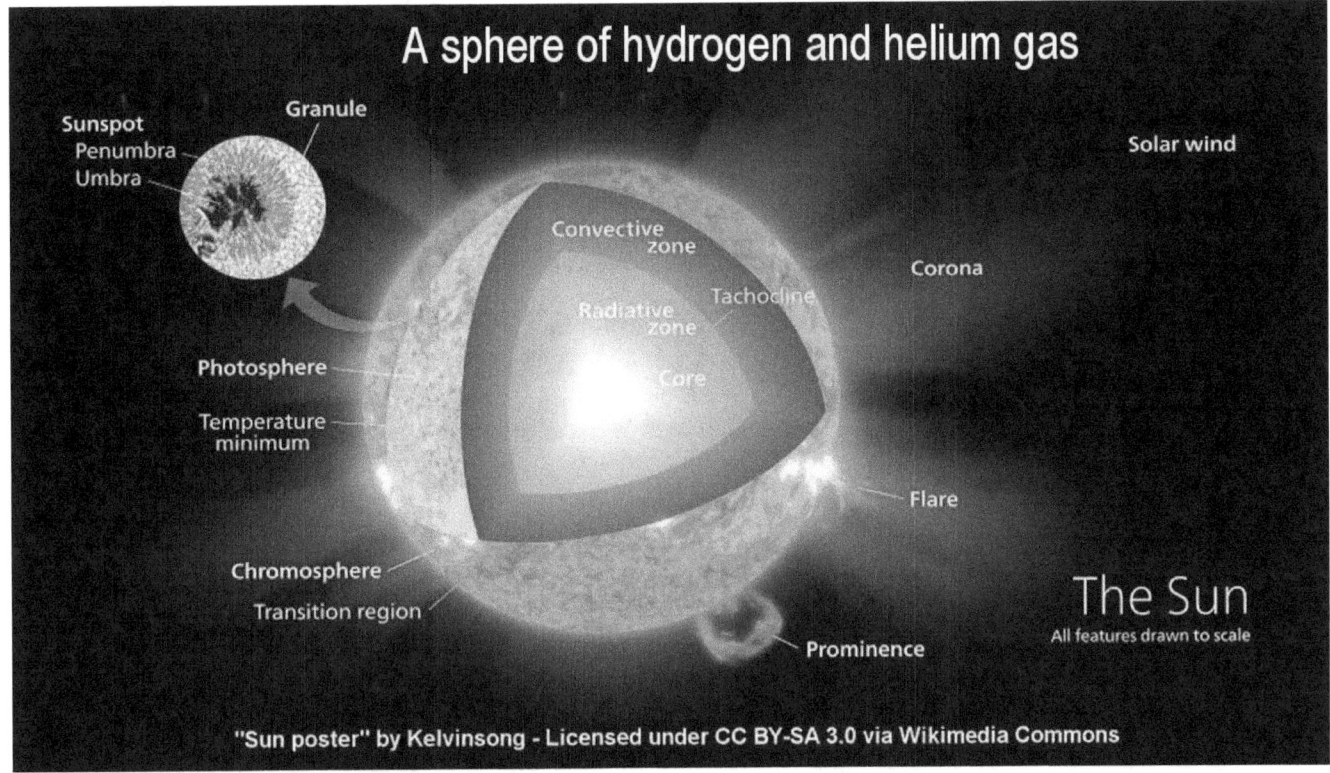

Shortly thereafter, the Hydrogen-Fusion Sun doctrine was invented, of a Sun that is its own master.

And a bit later, the Big Bang doctrine

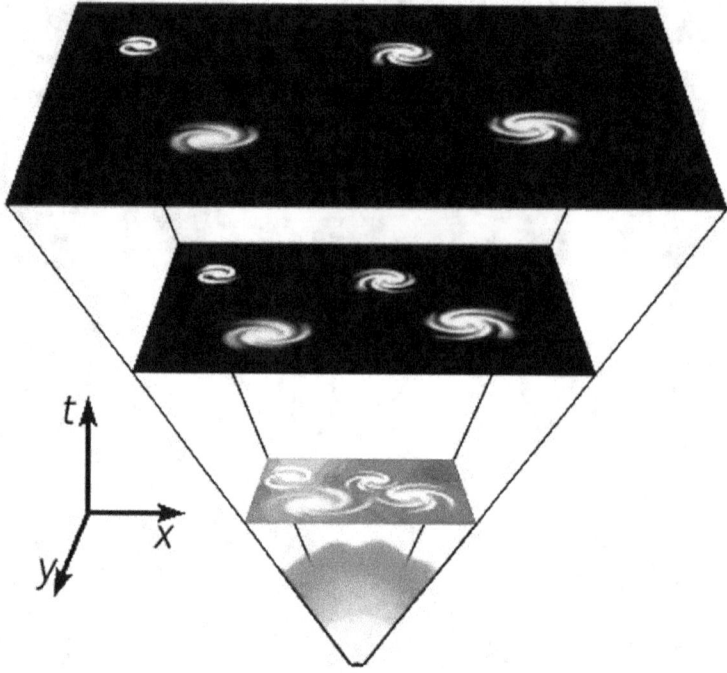

And a bit later, the Big Bang doctrine was invented to support the hydrogen gas Sun. The Big Bang was said to have created the hydrogen for the Sun.

The 'Invariable Solar Constant'

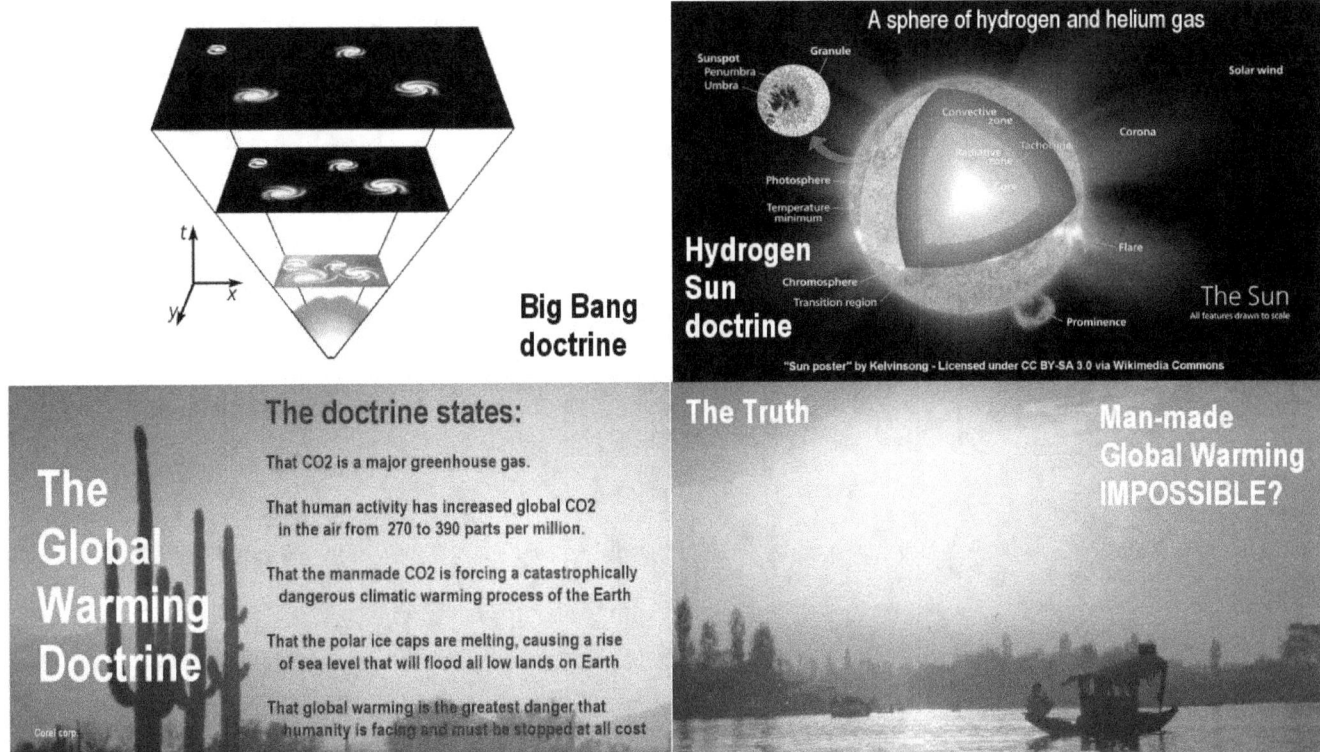

With the two-pronged 'devil's fork' established, a foundation was set up for the doctrine on which the 'Invariable Solar Constant', as it was called, could be 'sold' to society. The doctrine so established, effectively turned the lights out in the realm of astrophysical science. Climate Change was blamed on humanity.

The Ice Ages were rendered by the doctrine of the Constant Sun, a purely mechanistic phenomenon.

The 'devils fork' came out of the 1920s

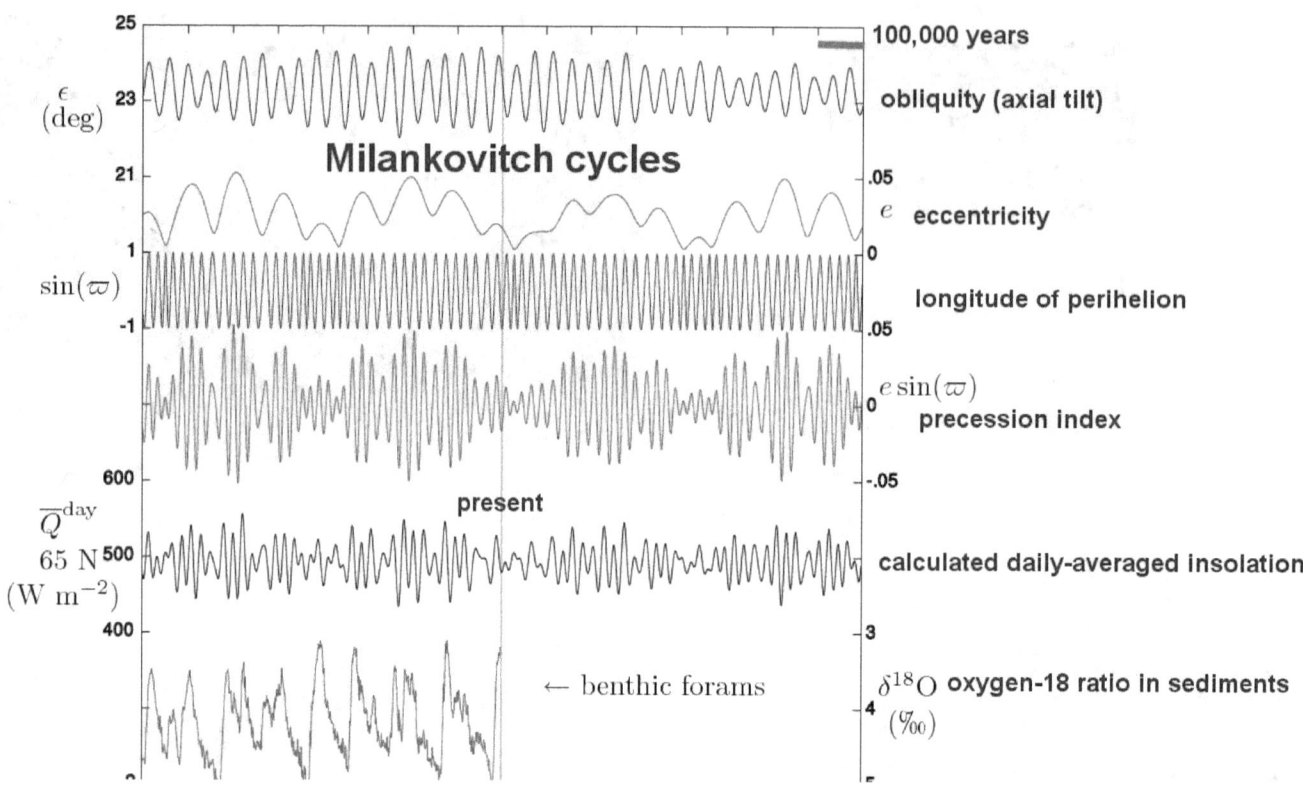

As I said before, the Ice Ages were deemed to be the result of numerous minute variables of the orbit of the Earth, all interacting, all measured in tens of thousands of years.

The 'devils fork' came out of the 1920s, during the decade of the great debate of how to deal with science. The Manmade Global Warming project was added 50 years later, in the 1970s.

With the devils fork worshipped in astrophysics as a god, the Ice Age subject was rendered to be of no concern to anyone. The subject was closed.

If the Ice Age 'cat' was to slip out

The Elite's great fear evidently was, that if the Ice Age 'cat' was to slip out of the bag, especially the fact that the next Ice Age is imminent, then worldwide economic development that the recognition would spark, would end the power of the noble Elite forever, which has no foundation to exist in a developing world.

Consequently, the scientific aspects of the Ice Age dynamics have remained artfully hidden from society in order that the imperial oligarchic Elite of the world would be able to continue to exist as in olden times, powered by slavery, looting, and stealing, which accumulates wealth and fosters impotence in society, and which also keeps the world in turmoil with wars and famines, towards the self-destruction of society.

Ice Age recognition kept out of sight

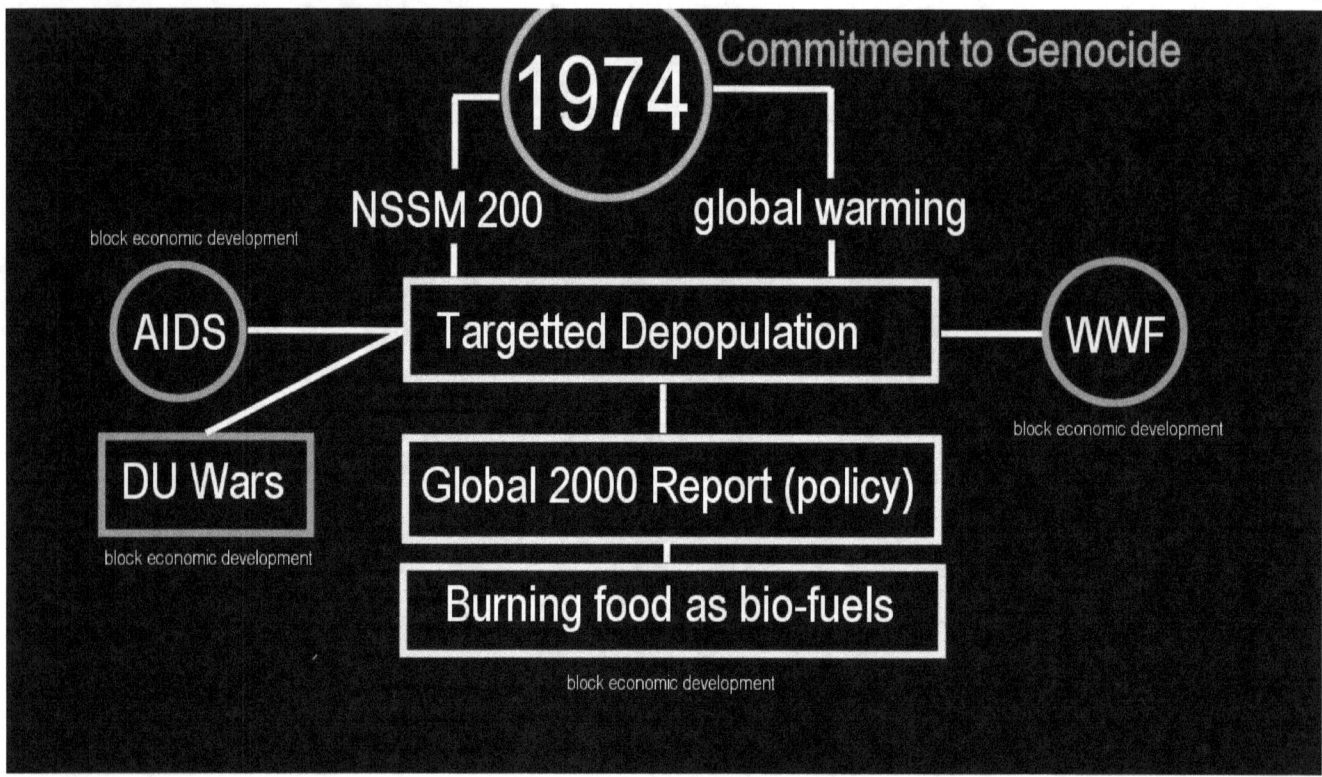

The critical Ice Age recognition is therefore kept out of sight, on this basis, ridiculed and denied. Obviously it was understood by the elite that whenever the Ice Age recognition came to the surface, it would have to be squashed by all means possible. This resulted in the big project in the 1970s with which the Manmade Global Warming doctrine was launched. The growing scientific concern at the time, about the coming Ice Age, that had surfaced at the time, was quickly pushed back out of sight with the Manmade Global Warming doctrine, and was kept out of sight, successfully, to the present day.

➢ **What forces cause an Ice Age?**

What are the forces that cause an Ice Age?
How does the Sun change?

What dynamics are at play?
What is society encouraged NOT to see?

What forces cause an Ice Age? How does the Sun change? What dynamics are at play? What is society encouraged NOT to see?

Recognition of the Sun as a sphere of plasma

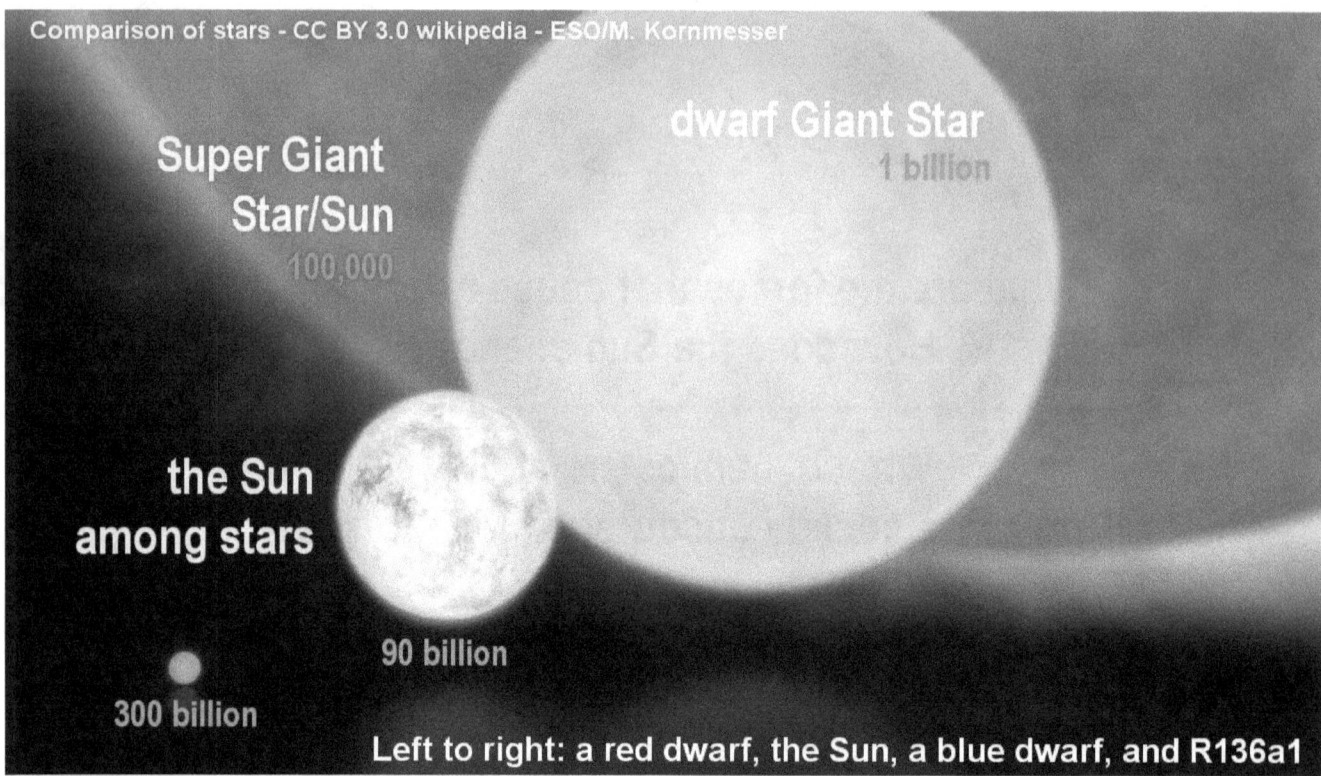

An answer began to develop with the discovery of a natural principle in electrodynamics that enables the recognition of the Sun as a sphere of plasma that responds to electric-force principles, instead of being a sphere of atomic gas that is ruled by gravity, because a sphere of atomic gas of the size of the Sun, or even larger spheres, cannot actually exist. Such spheres would far exceed the gas compression limit.

All the known parameters about the Sun

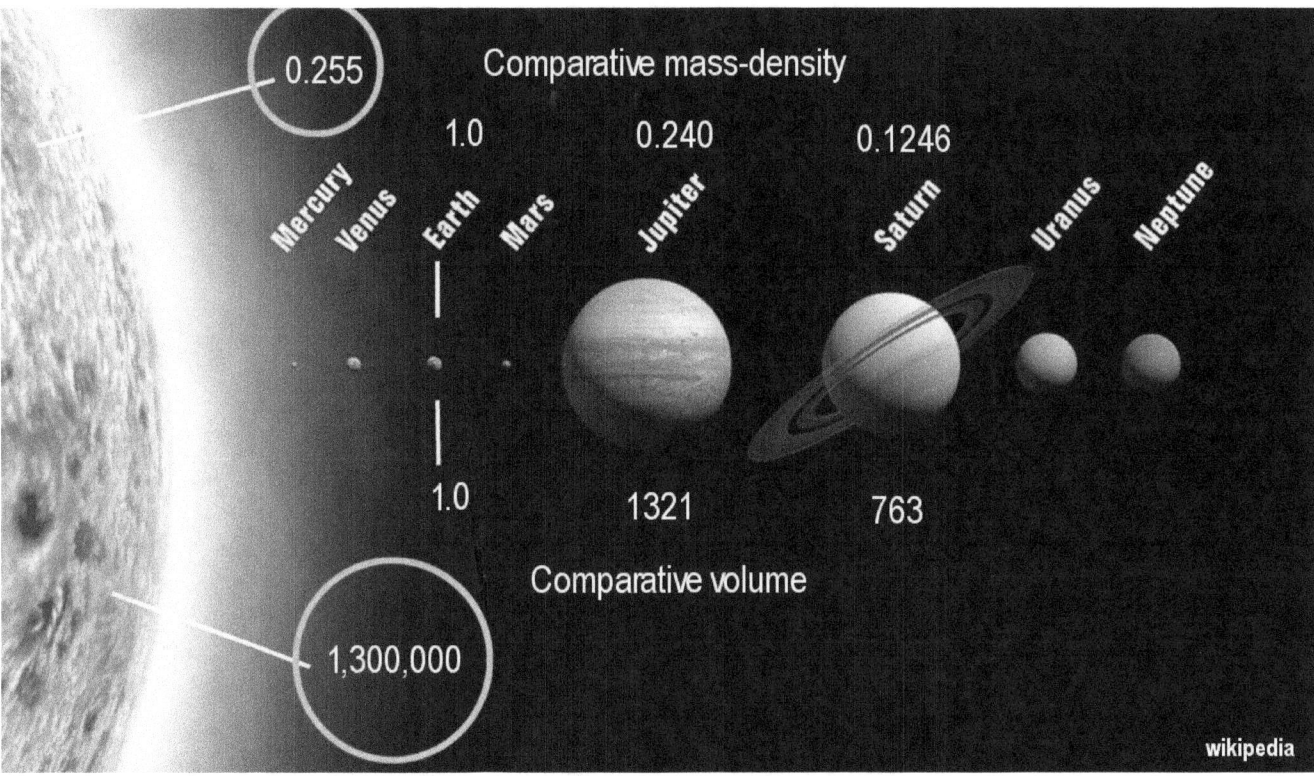

And if a gas sphere of the size of the Sun could magically exist, it would be a thousand times heavier than it is known to be. But as it is, all the known parameters about the Sun, none of which fit a gas sphere, match the characteristics of a plasma Sun.

A Plasma Sun is an extremely variable star

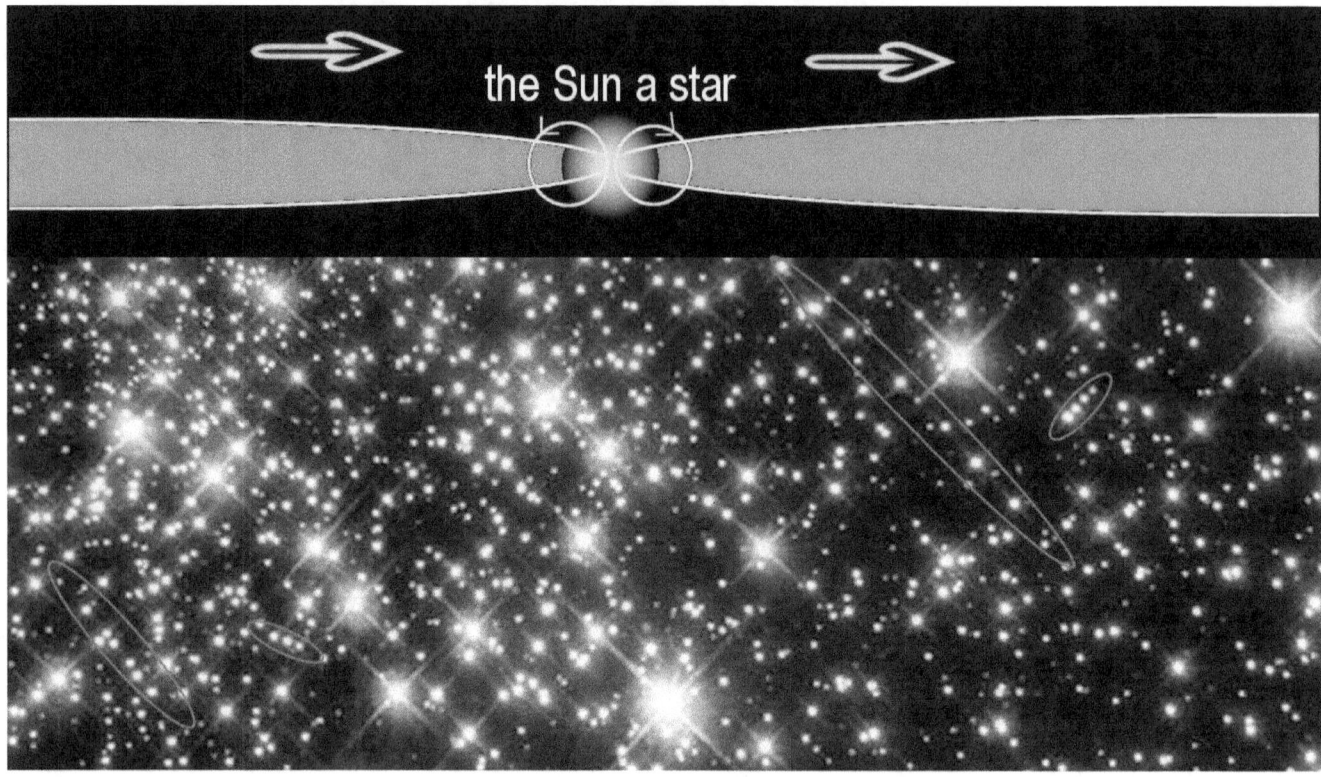

A Plasma Sun is an extremely variable star, by its very operating principle. It is not its own master, but merely responds to changing external conditions. And it responds quickly. Some of its responses, now that the solar system is getting weaker, cause enormously large effects on the Earth, such as flooding, drought, blizzards, freezing, and so on.

Since the Sun is externally powered

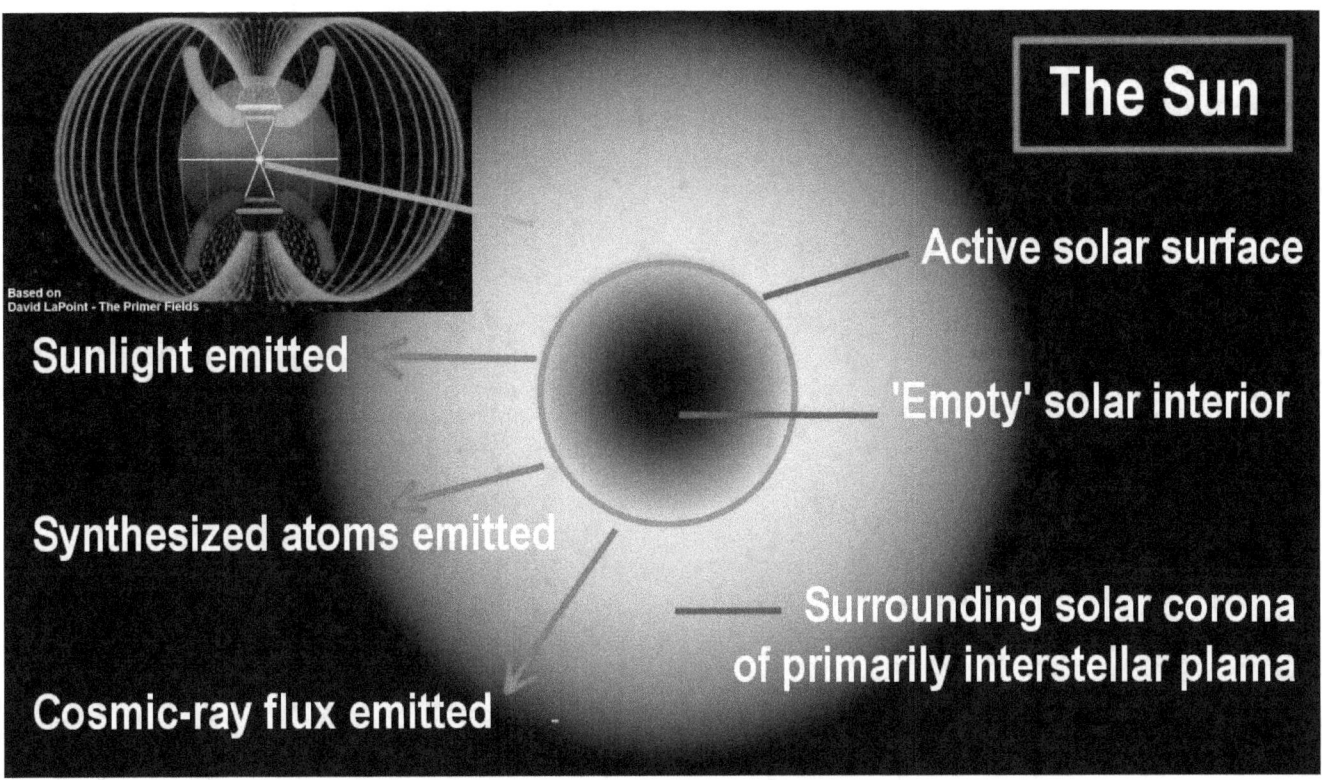

Since the Sun is externally powered, with plasma interaction on its surface, the obvious question is: How is this possible?

A sink effect is needed

A number of theories were developed. One of these is the Electric Sun theory. The theory seems reasonable, except for a fundamental impossibility. The Electric Sun theory has electrons flowing into the sun that generate heat and light in an electric-arc type interaction. But this isn't possible on a continuous basis. The electrons flowing into the Sun would pile up and clog the reaction process. A sink effect is needed, like an open faucet. The open faucet enables the water to flow, it enables the dynamics of flow. Every dynamic flow process requires both a source and a sink. The Electric Sun theory doesn't incorporate the critical continuous sink effect. But the Plasma Sun does.

The Plasma Sun, in comparison, is powered by inflowing streams of plasma that are bound up on the surface of the Sun into atomic elements that are electrically neutral and flow away with the 'wind.' The atomic synthesis provides the sink effect. The plasma that flows into the atomic synthesis no longer exists in the electrically organized plasma environment as if it had exited the universe. In this manner the Plasma Sun 'consumes' plasma. The atomic synthesis is the open faucet, in this comparison. Its plasma consumption enables interstellar plasma streams, both to form, and to flow.

By being electrically neutral, the created atoms are no longer a part of the landscape of electric forces. They are free to flow into space, typically with the solar 'wind,' and they do so massively. All the planets in the solar system were created in this manner over time. The plasma-consuming sink effect keeps the plasma streams in motion, flowing into the Sun, for ever and ever.

The Plasma Sun is largely hollow inside

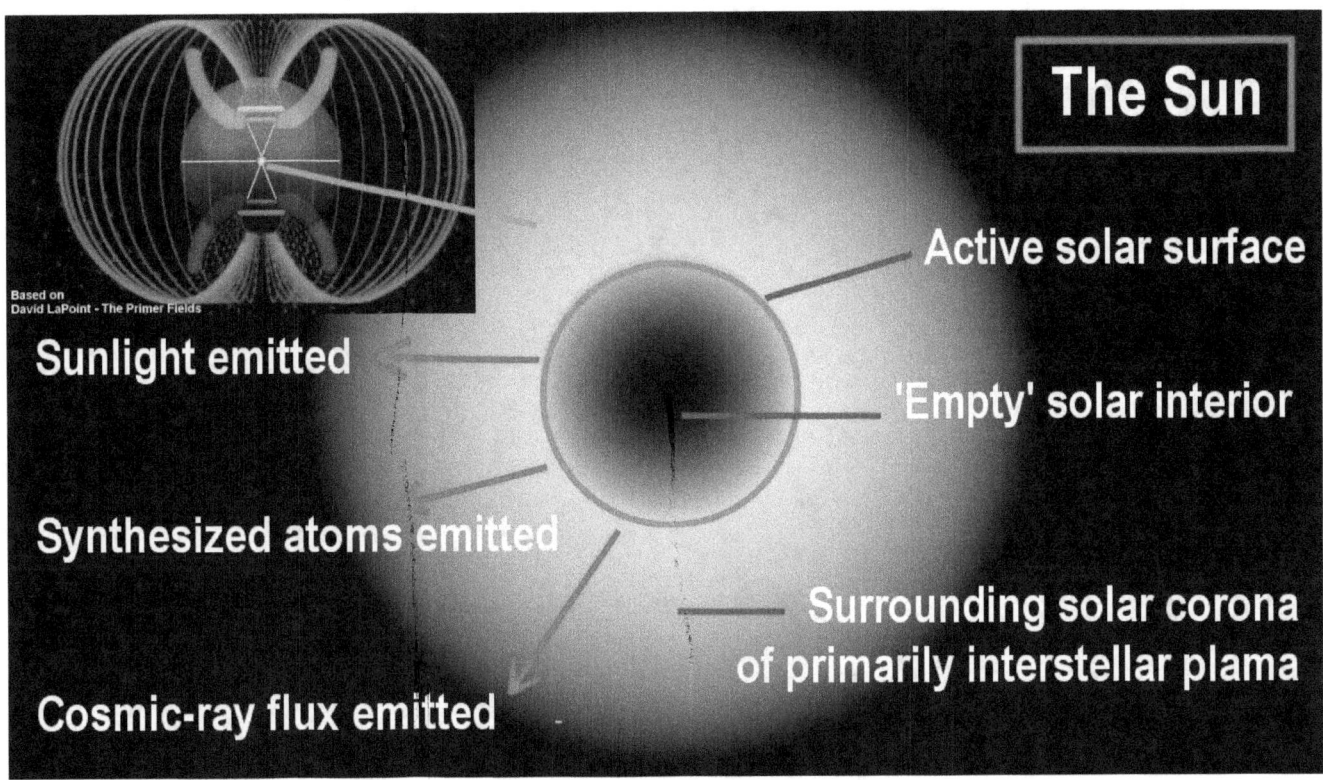

The Plasma Sun is a sphere of plasma that is largely hollow inside.

In extremely large spheres of plasma

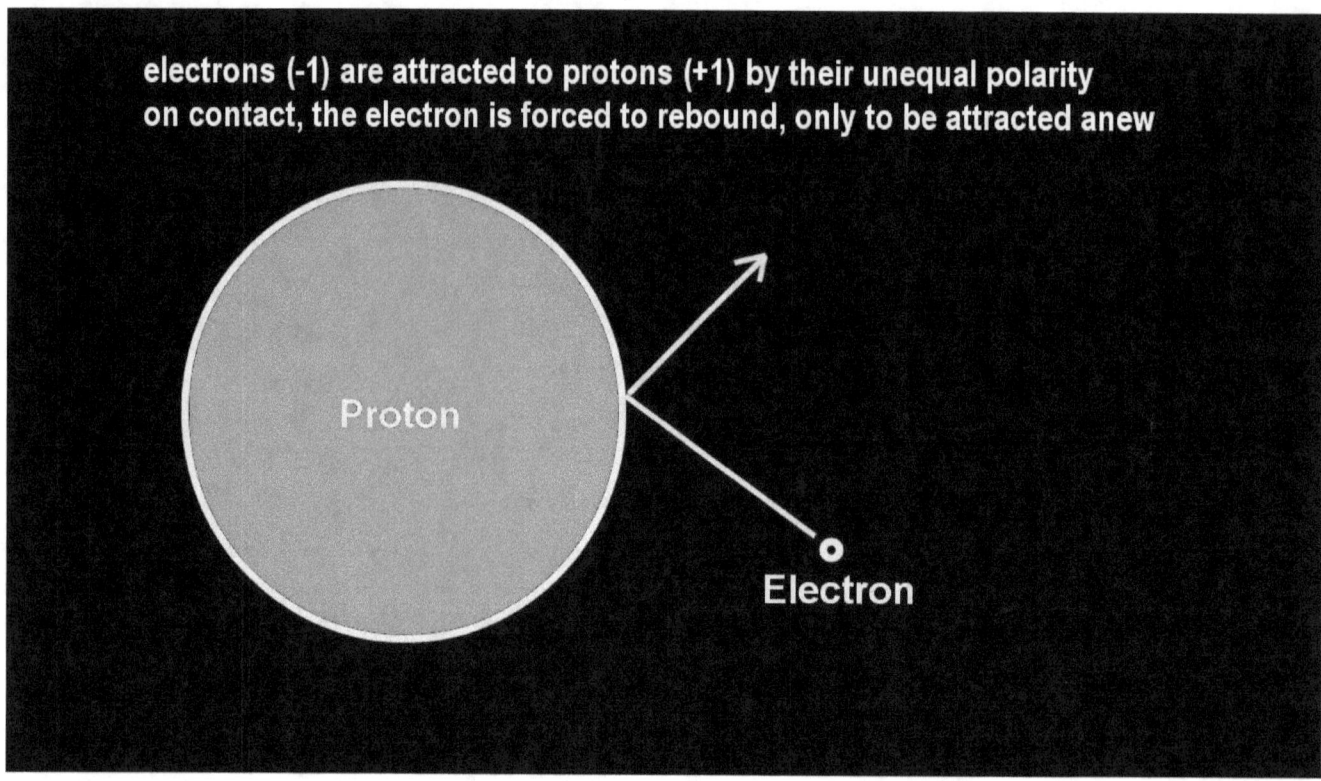

In extremely large spheres of plasma, gravity begins to play a role, which affects the dynamics of the plasma that is always in motion. Plasma is made up of protons and electrons, which carry a positive and negative electric charge respectively. Particles of opposite charge attract one another with the electric force, and those of like charge repel one another. By this electric principle the negative electrons are drawn to the positive protons with the electric force that is one of the strongest forces in the universe. But before the electron can latch itself onto a proton, an even stronger force repels the electron again, only to be attracted anew.

With the electrons being a thousand times smaller than the protons, the electrons become drawn into an endless swarming dance around the protons. The swarming effect also enables the protons to be more densely packed to each other. However, in very large spheres of plasma where gravity also plays a role, the lighter electrons tend to swarm away from the center of gravity, towards the surface. The migration of the swarming electrons towards the surface, enables the protons at the core to repel each other more strongly.

A plasma sphere has the greatest mass density at its surface

The end result is that a plasma sphere has the least mass density at its center, and the greatest mass density at its surface, which is the very opposite characteristic of a gas sphere. Enormously large plasma structures can thereby form, which in the extreme become largely hollow shells. A sun needs to have an extremely large surface area, in order to be able to emit large volumes of radiated energy. In addition, a plasma sphere also has an extremely high electron density at its surface. The density is so great that it attracts interstellar plasma to it and strongly reacts with it. It is in principle impossible for a large plasma sphere not to be a sun.

The pinch effect

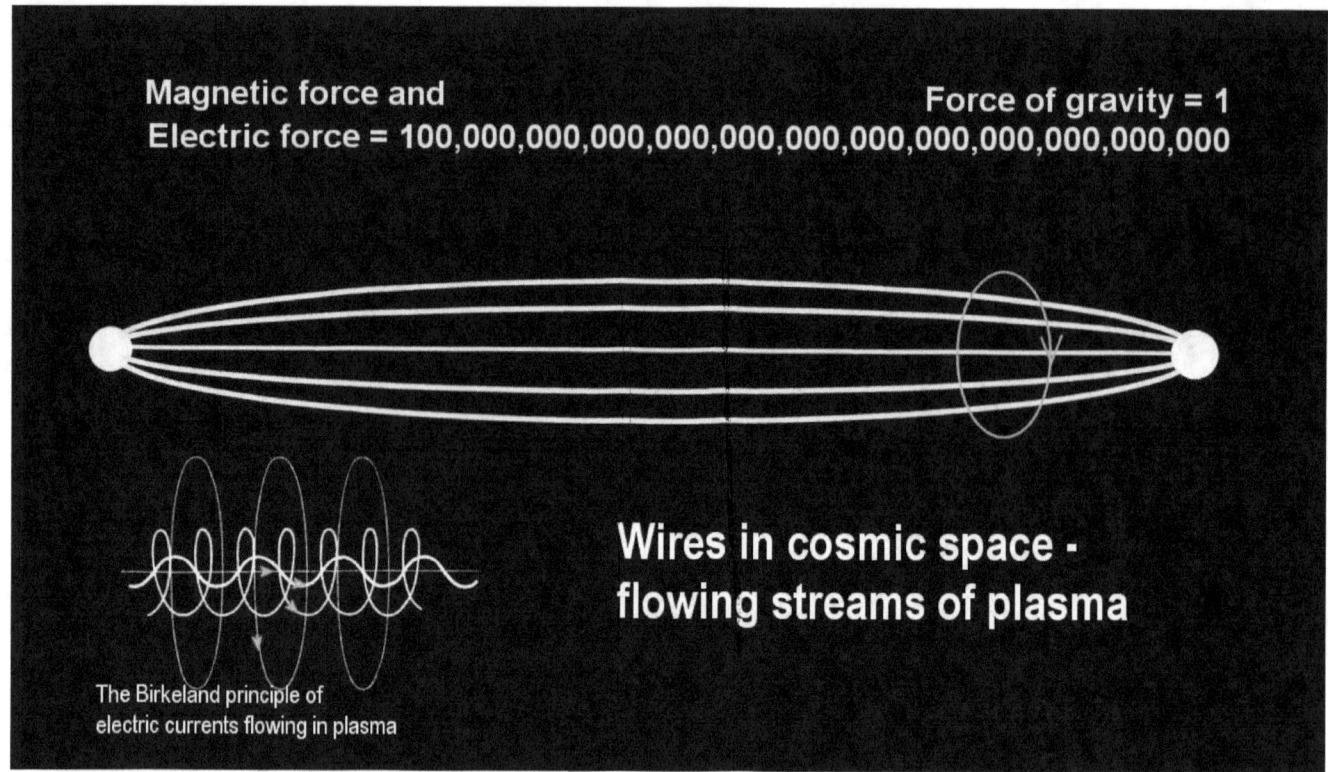

But the process doesn't stop at this stage. When large volumes of plasma flow in cosmic space, the magnetic fields that the movement of electric particles create, pinch the moving plasma streams together into ever-tighter concentrations. The pinch effect, of course increases the magnetic concentration likewise, which in turn increases the pinching effect evermore.

Here things become interesting.

High-energy discharge experiments

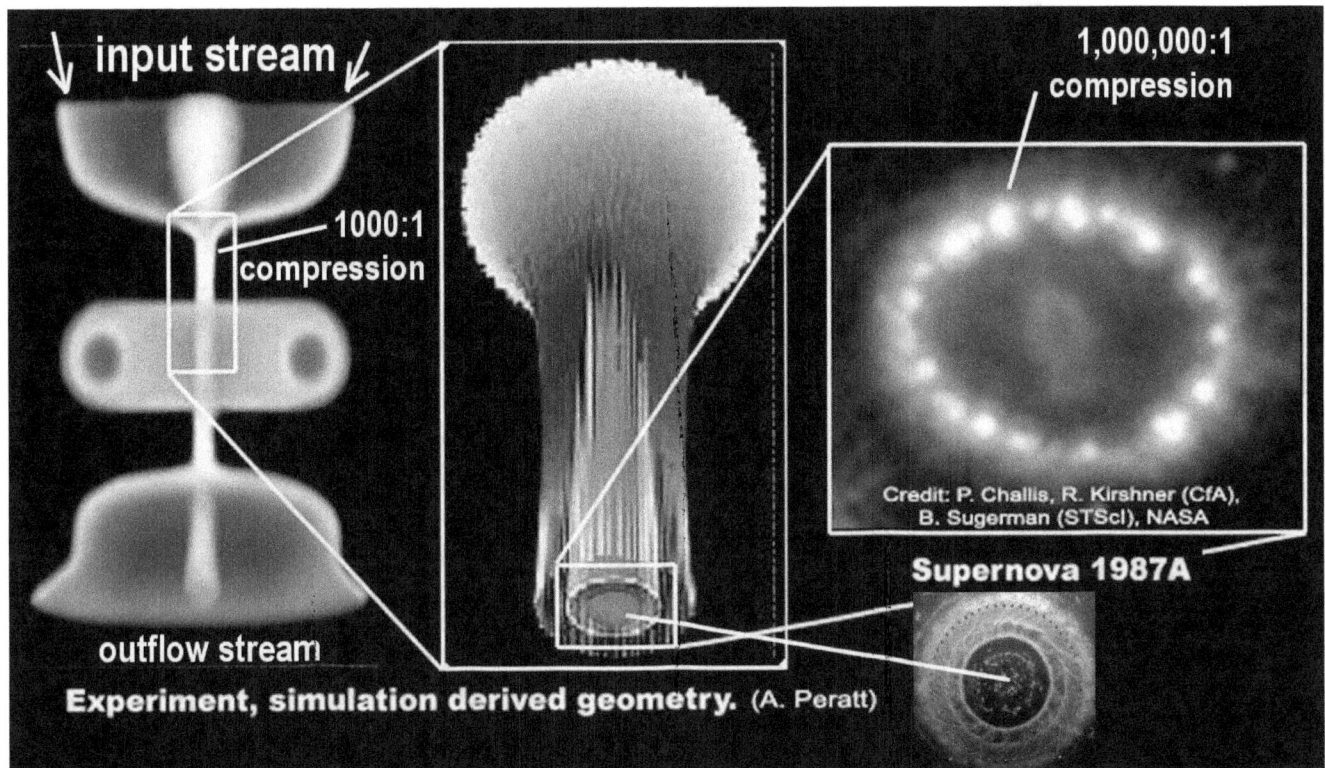

Researchers at the Los Alamos National Laboratory had conducted a series of high-energy discharge experiments, in order to discover what physical shapes result when the magnetic pinching effect becomes so extreme that the plasma particles literally flow into each other from all sides, with nowhere to go.

In such cases different physical principles come into play that force the plasma and its magnetic field to fold backwards into a dome-like structure where the plasma becomes super-concentrated until the pressure becomes intense enough to break out of the confinement into a thin stream that forms a ring-like structure of 56 rotating filaments of hyper-dense plasma.

In the experiment, the super-intense magnetic field, also generated a toroidal structure around the thin stream. The energy dissipation that results, enables the now weakened plasma stream, in the experiment, to expand again in the reverse of the process.

Toroidial structures are visible

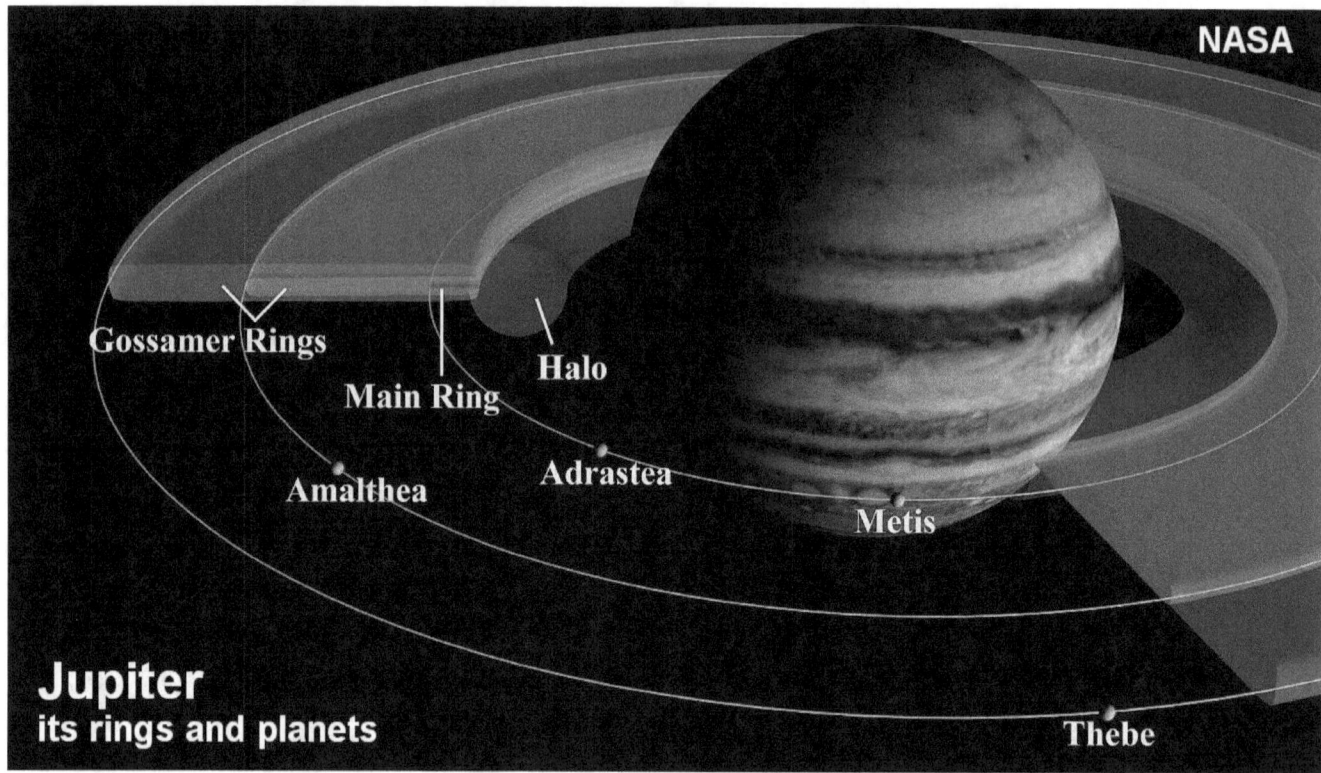

In a few cases in cosmic space, toroidial structures are visible. We see an example in the form of a halo around the equator of Jupiter. The rings extend away from the planet along the planet's ecliptic.

In the case of the Sun

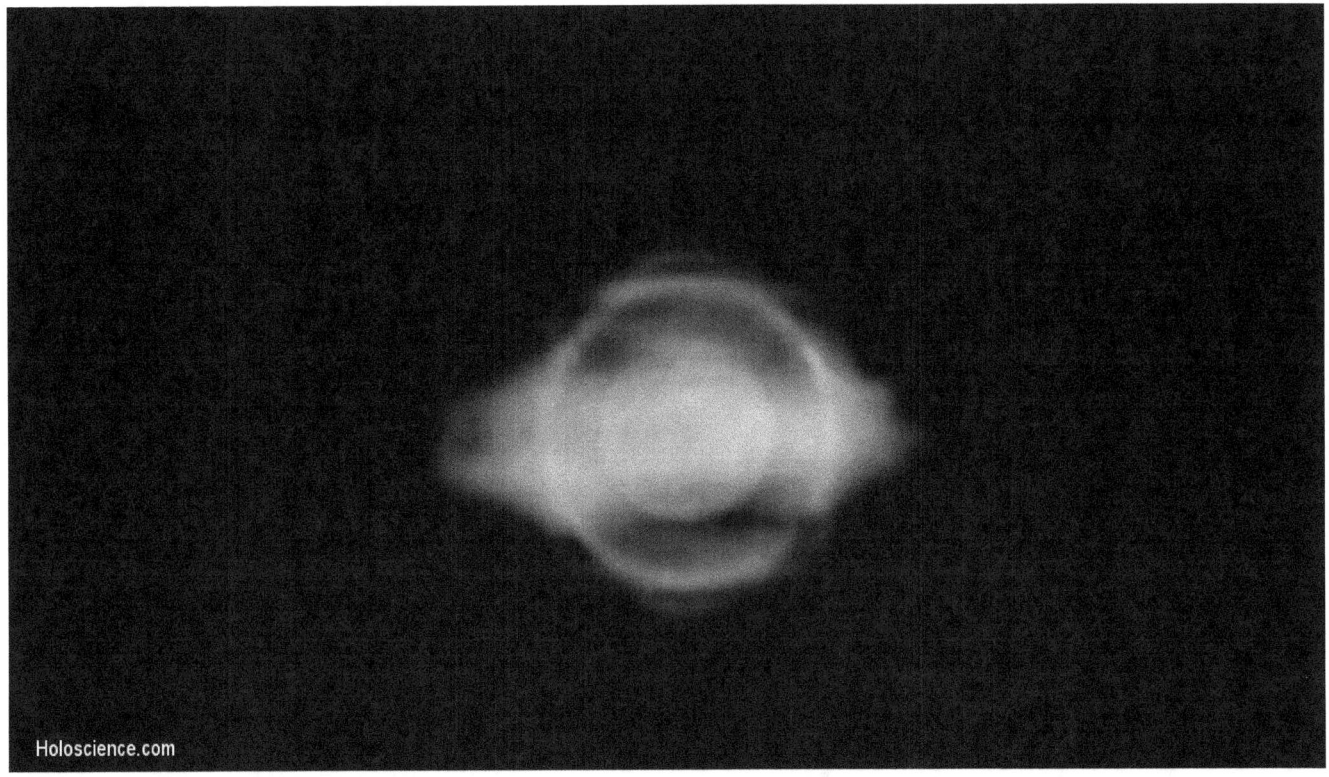

The toroidial structure also appears to have been seen faintly apparent around the Sun, as is shown here; though the image may be an artist's impression.

In the case of the Sun, it is the heliospheric current sheet that extends outward from it, like Jupiter's rings. The current sheet flowing from the Sun, extends for 15 billion kilometers, all the way to the edge of the heliosphere.

The heliospheric current sheet is aligned

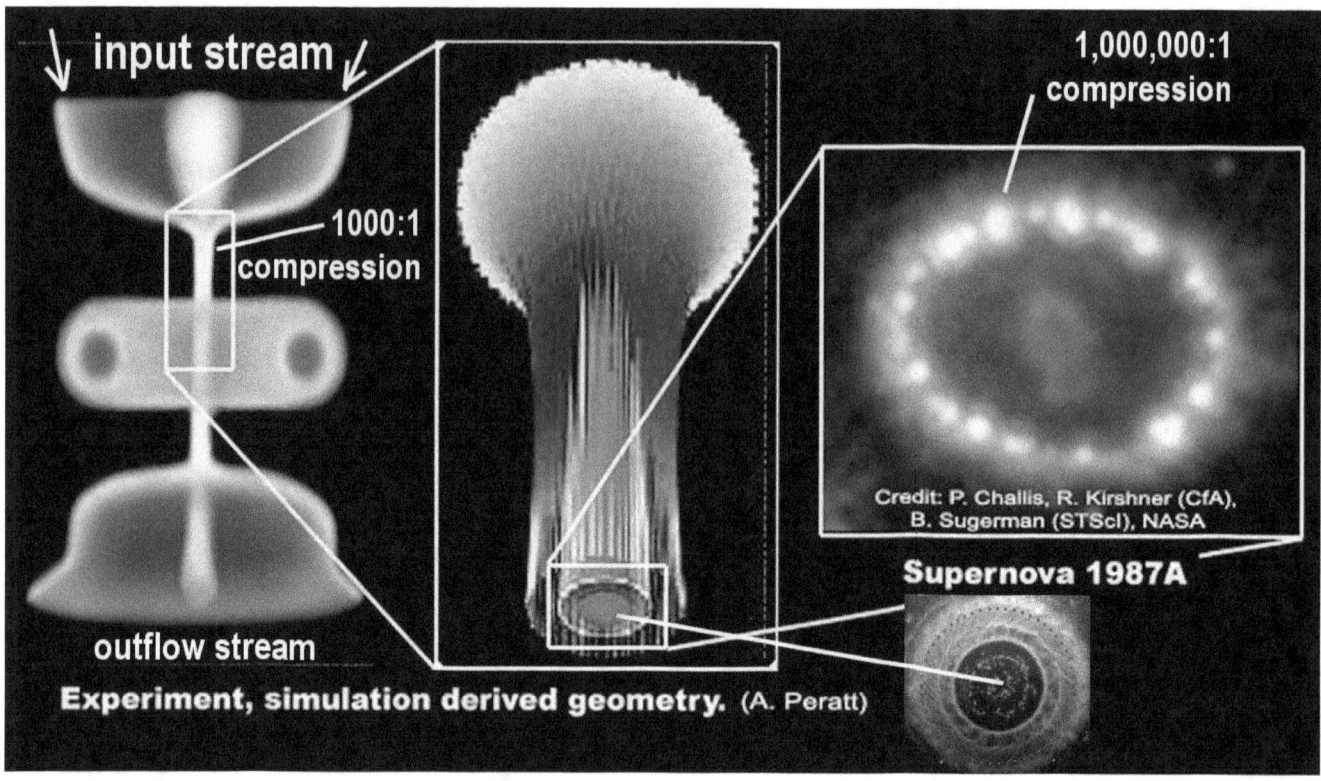

Close to the plasma stream, the heliospheric current sheet is aligned with the equator of the Sun, in the place of the toroidial ring, and further out, it becomes aligned with the ecliptic of the solar system.

../scrp/tngb018.htm

In the case of the Sun

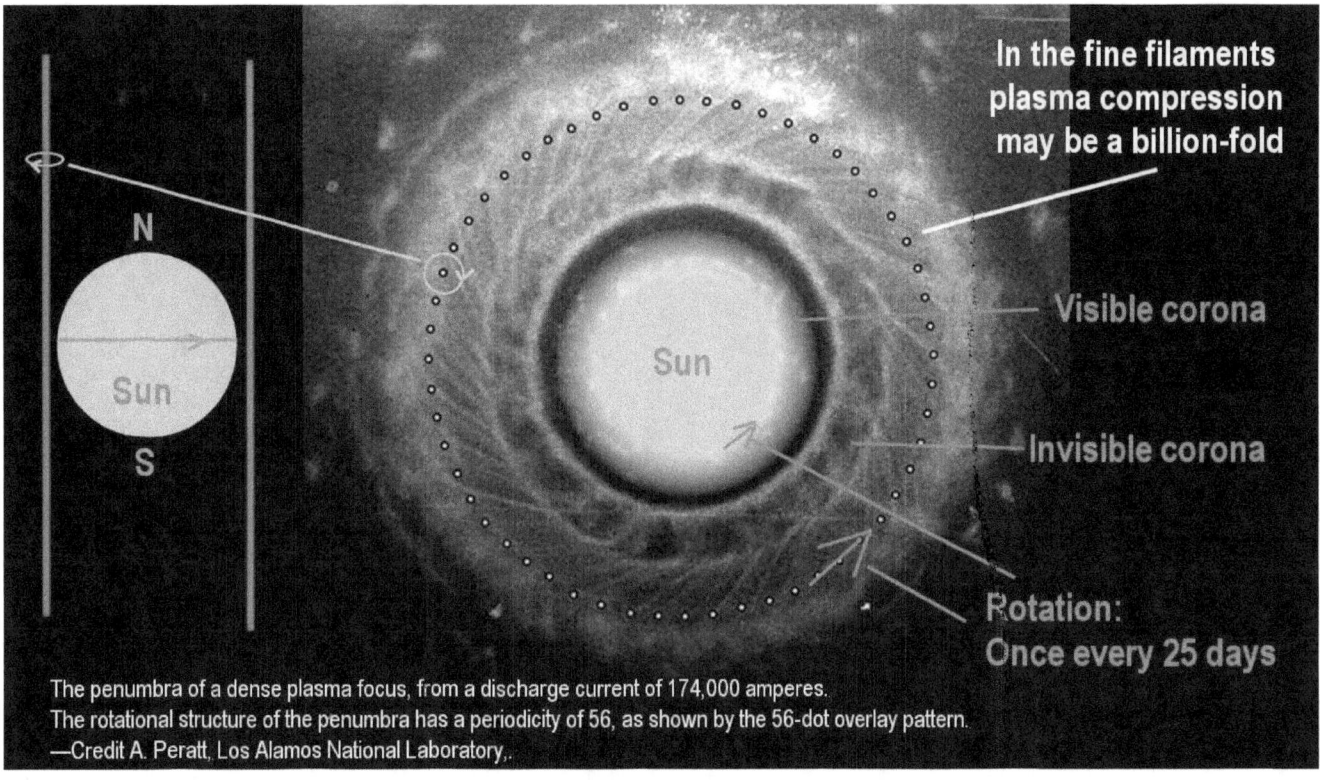

The penumbra of a dense plasma focus, from a discharge current of 174,000 amperes.
The rotational structure of the penumbra has a periodicity of 56, as shown by the 56-dot overlay pattern.
—Credit A. Peratt, Los Alamos National Laboratory,.

In the case of the Sun, the Sun itself would likely be located within the hyper concentrated plasma stream that extends between the complimentary magnetic structures, as was replicated in laboratory experiments.

In large plasma streams, the Primer Fields

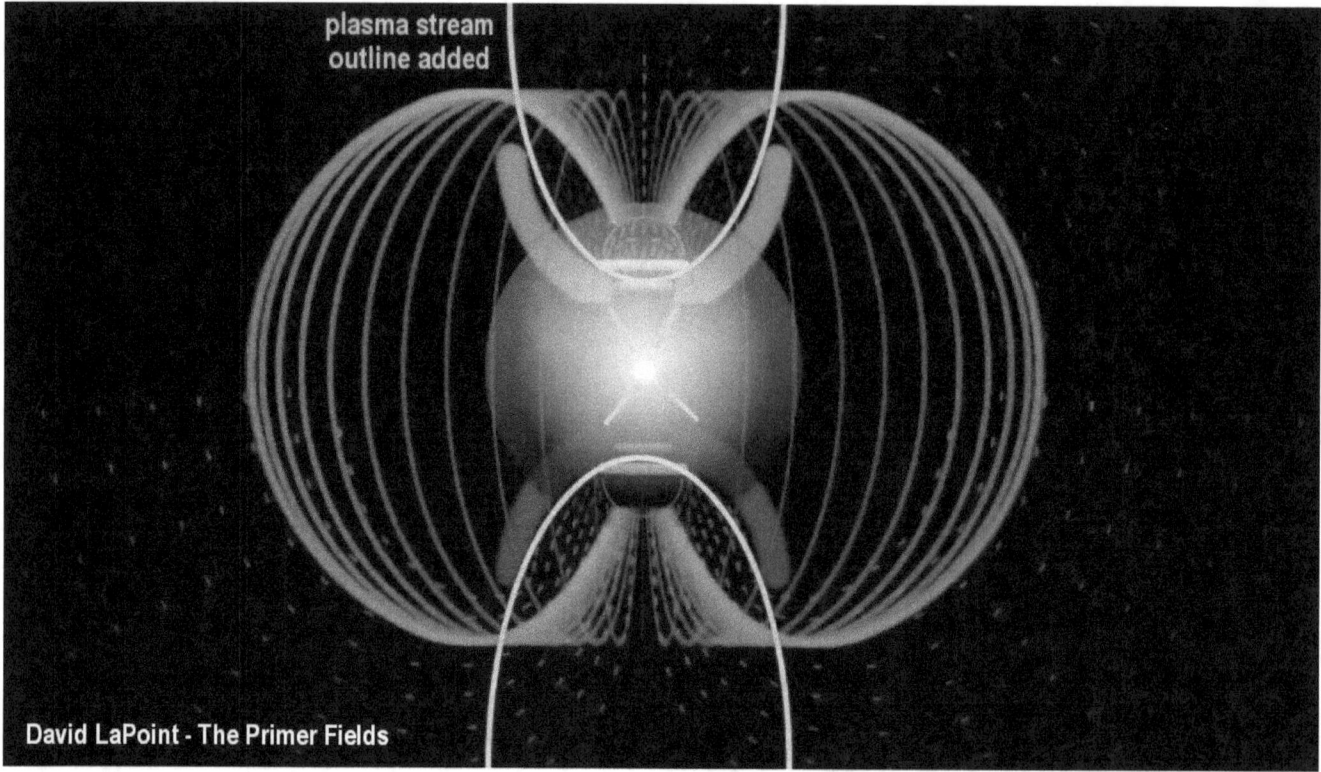

The plasma researcher David LaPoint termed the dynamically self-forming magnetic structures in large plasma streams, the Primer Fields.

In space, the Primer Fields form dynamically and concentrate plasma around a sun.

In order to explore the magnetic fields' characteristic in the small, David LaPoint manufactured himself a set of permanent magnets of the shape that the natural dynamics form.

In principle the powering of a Sun

With the replicas, he was able to replicate in principle the powering of a Sun, and explore features related to the process, all with a table-top experiment.

Experiments, by LaPoint and Peratt

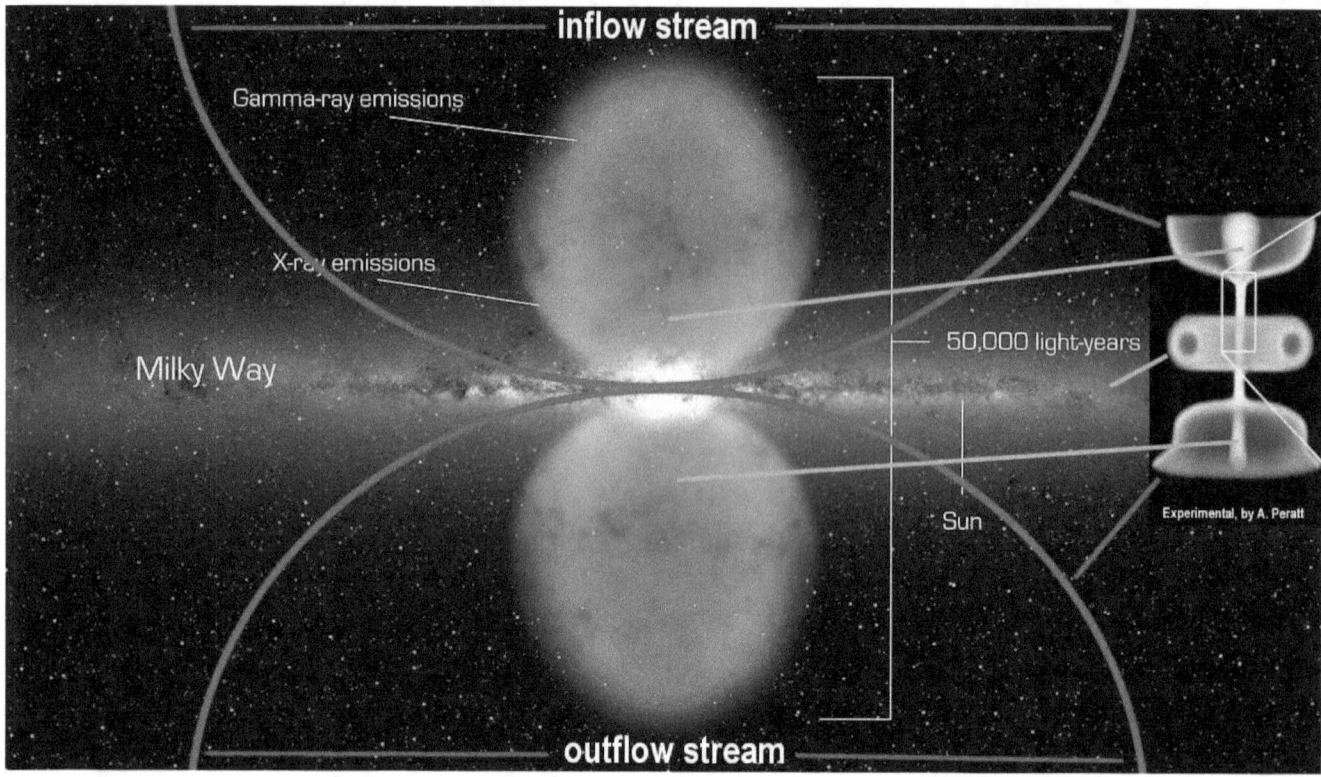

The features that were discovered with the two diverse types of laboratory experiments, by LaPoint and Peratt, have been subsequently discovered by NASA in space, both on the stellar scale, and also in its very large form on the galactic scale, that is shown here.

NASA's Ulysses satellite that had orbited the Sun

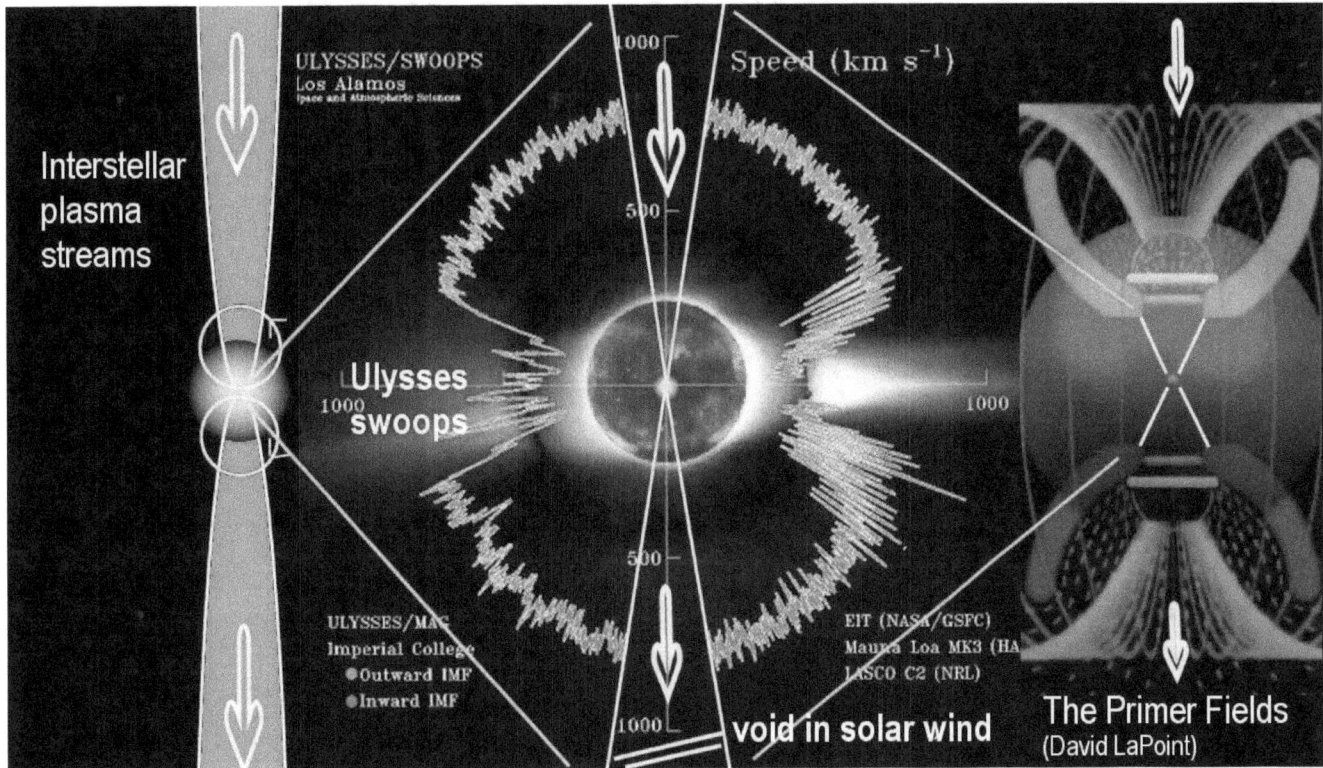

The evidence that large plasma-stream structures do exist in cosmic space, with plasma being focused onto our Sun, has been brought home by NASA's Ulysses satellite that had orbited the Sun in a polar orbit for 16 years.

The Ulysses satellite had measured a sharply delineated void within the sphere of the out-flowing solar wind and also magnetic fields. NASA has measured the void over the polar regions of the Sun, in both hemispheres, consistently, at every orbit, precisely where highly concentrated plasma is flowing unto the Sun according to the experimental models.

Verification in cosmic space

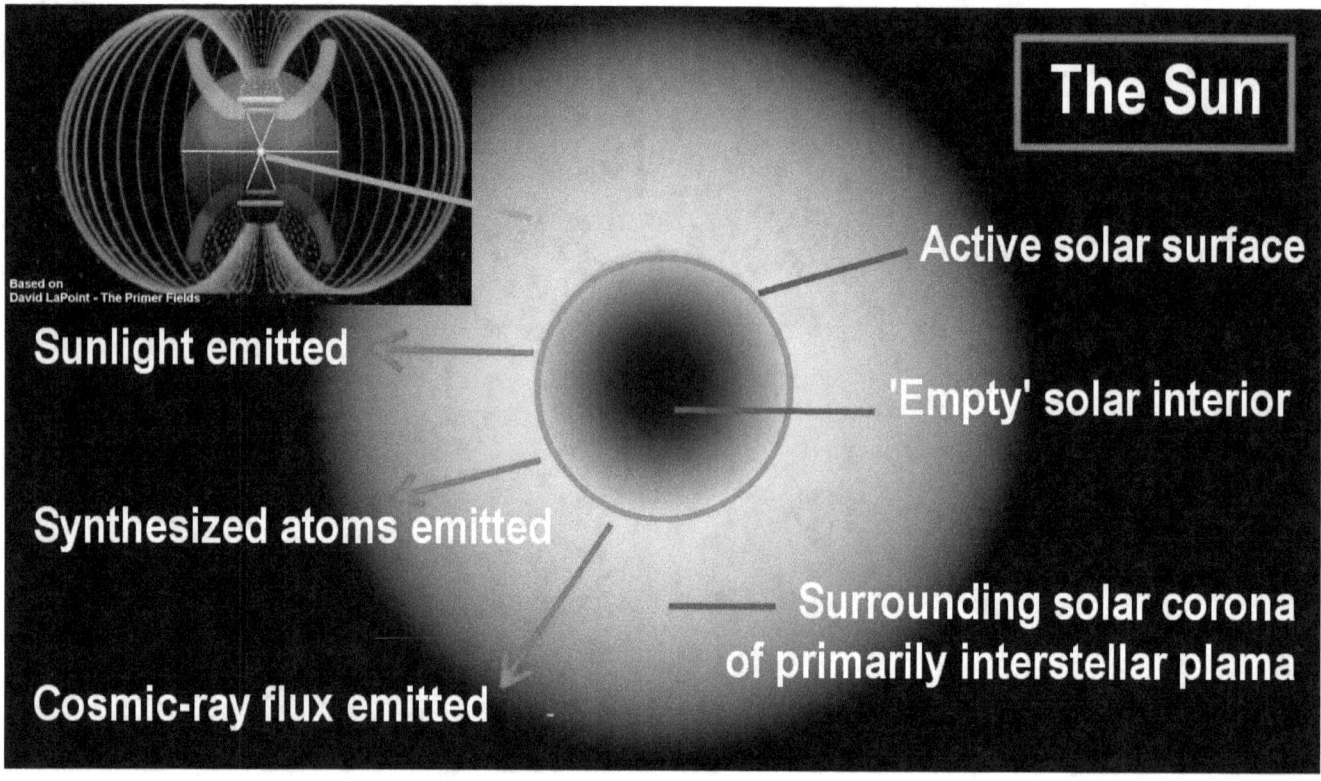

The verification in cosmic space, of the plasma stream that drives plasma unto the Sun, is, all by itself, worth all the efforts that went into planning, building, and operating the amazingly successful Ulysses science project that extended over 30 years from start to finish.

➤ **To 'see' the invisible**

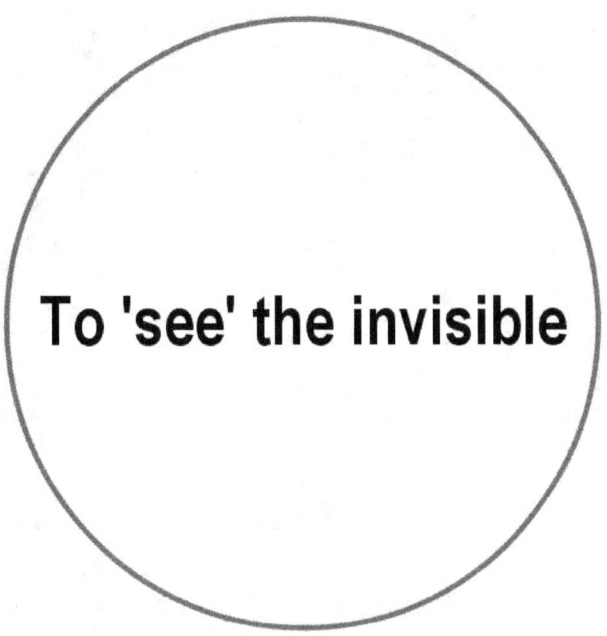

To 'see' the invisible

Ulysses, had made the invisible, 'visible.'

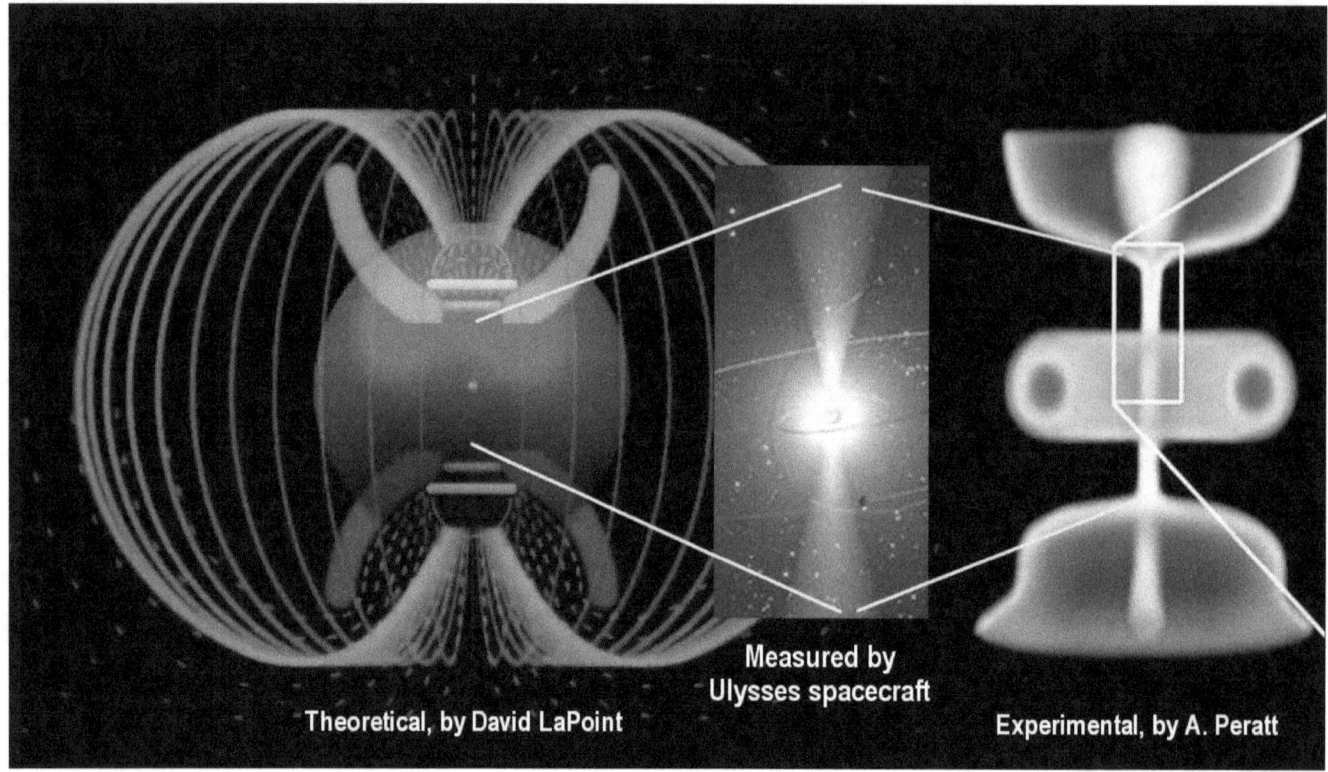

Ulysses, had made the invisible, 'visible.'

Plasma is not visible in space.

The protons in plasma are 100,000 times smaller than the smallest atoms, and the electrons are a thousand times smaller than that. But plasma can be 'seen' by its effects, such as by the void in the outflowing solar wind that Ulysses 'saw' over the Sun's poles.

Another 'visible' example of plasma streams

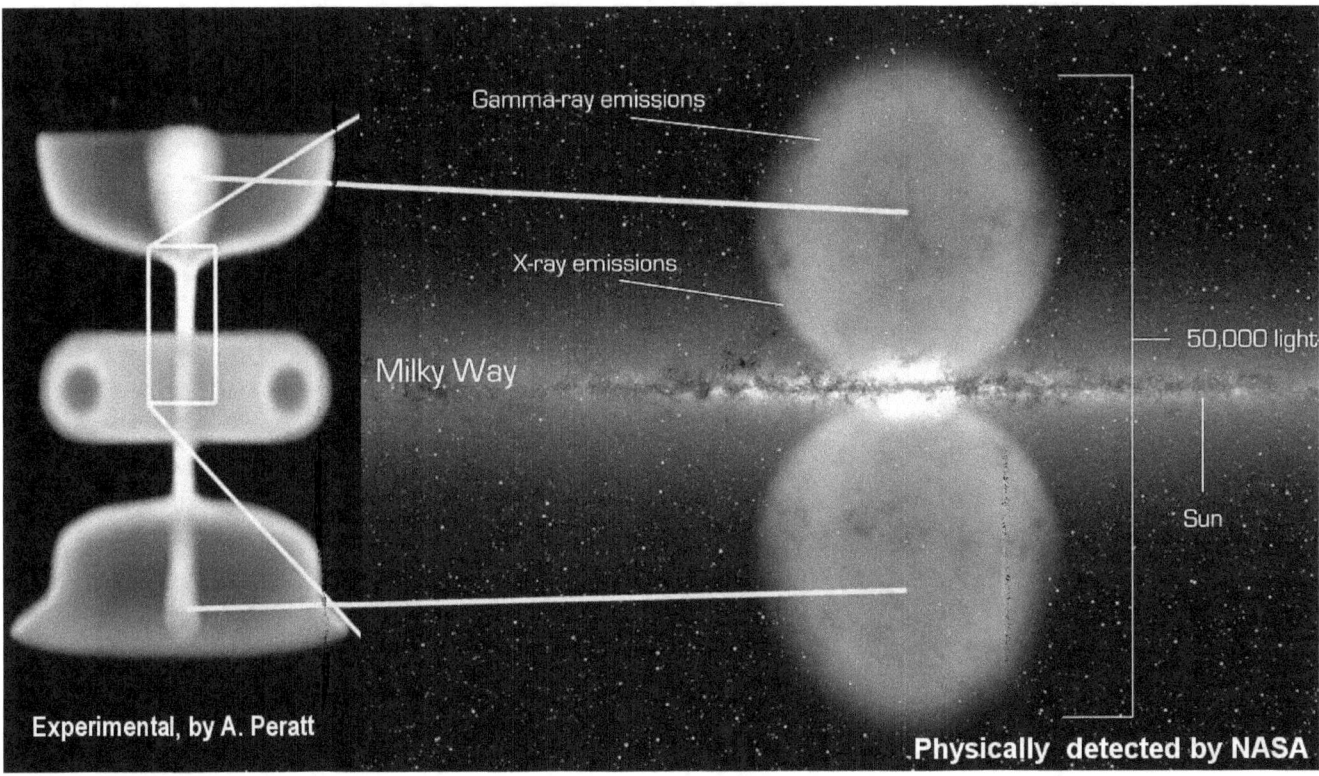

As I said before, another 'visible' example of plasma streams flowing in cosmic space, has been 'seen' above and below our galaxy. In this case the plasma structures were 'visible' in x-ray and gamma-ray light.

NASA saw two large confinement dome structures existing above and below the galactic disk. The structures are large. They extend for 25,000 light years in each direction.

Shape being that of a confinement dome

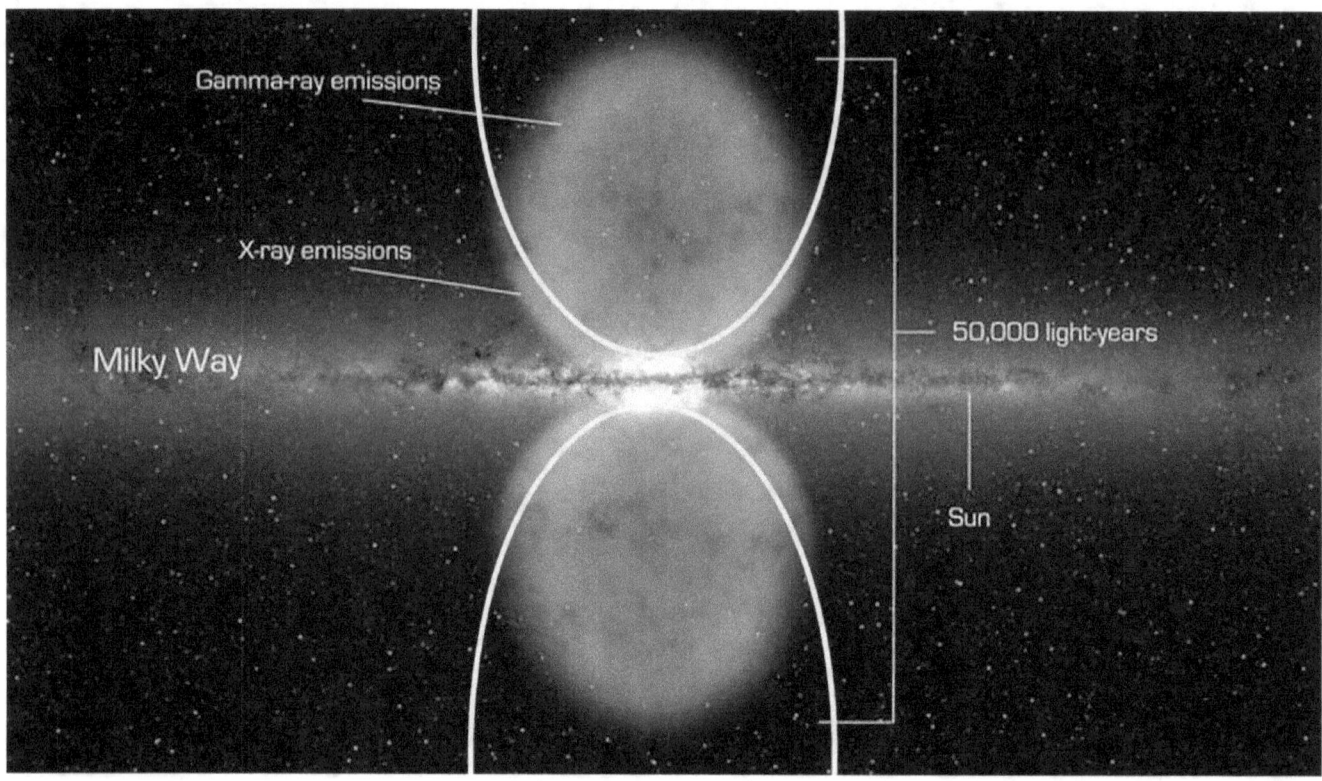

With their shape being that of a confinement dome, their presence illustrates that our entire galaxy is but a node-point structure between intergalactic plasma streams that power the galaxies and everything within them.

The galactic node point

In this case, the galactic node point would be between extremely long plasma streams with electric resonance features within the streams, measured in tens of millions of years.

In a hypothetical case for our galaxy

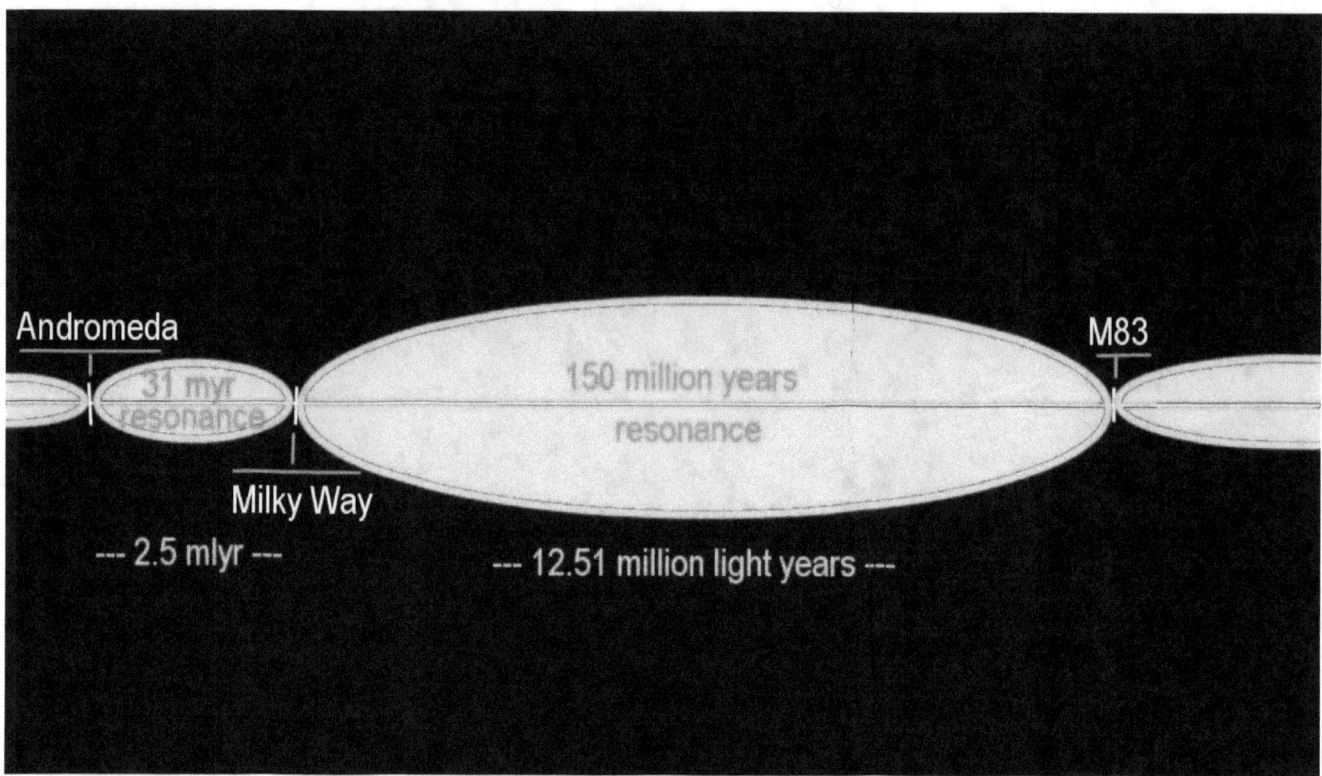

In a hypothetical case for our galaxy, the Milky Way Galaxy, one intergalactic plasma stream might connect with the Andromeda Galaxy, 2.5 million light years distant, with a resonance cycle of 31 million years. The other stream might connect with Galaxy M83, 12.5 million light years distant, with a 150 million years resonance.

All plasma streams have built-in resonance cycles, like the strings of a piano. In both cases the cycle times depend on the length and the density of the resonating structure. For long intergalactic plasma streams that span across millions of light years, plasma resonance times in the range of tens of millions of years, are well within the range of what one would expect.

But how can we make this type of theoretical plasma structure visible?

We have made it visible

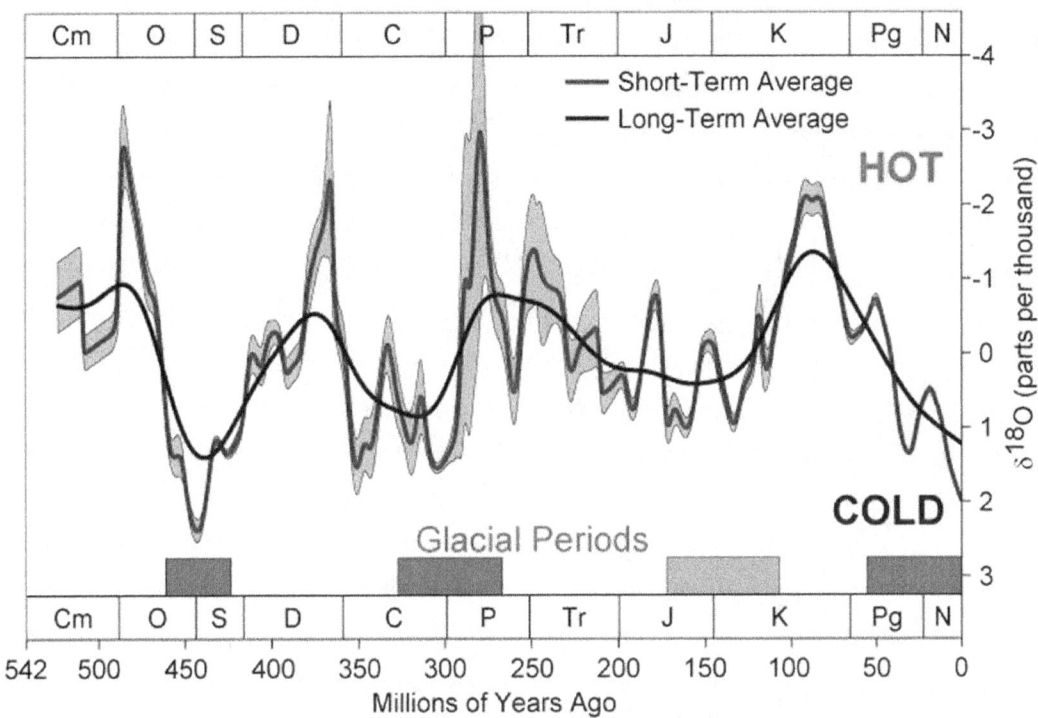

We have made it visible by measuring the heavy-oxygen ratio in sea shells from deep-sea sediments going back in time 500 million years. The heavy-oxygen ratio vary with climate temperatures, so that the measurements can be used to reconstruct the historic climate on the Earth. Now look what we find there. We find there two very-long climate cycles overlaid on each other, a 31 million years cycle, and a 150 million years cycle. That's how we can 'see' the resonances of the intergalactic plasma streams.

Galaxy is presently at the weakest point

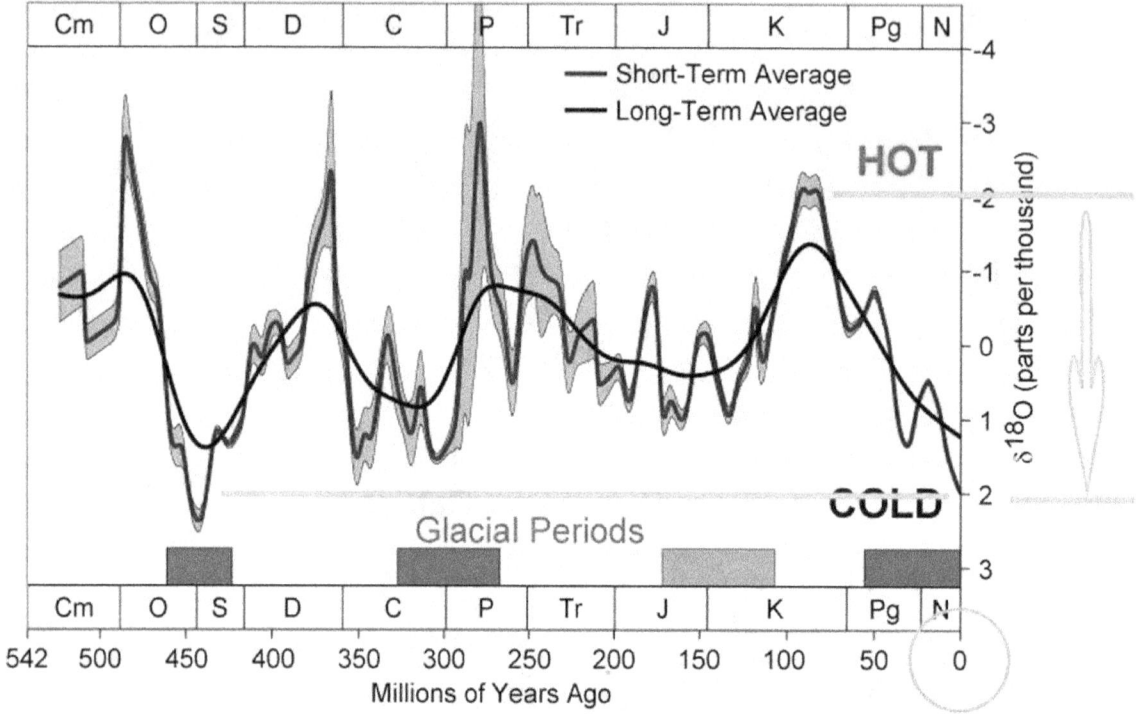

And note what else we find. We find that our galaxy is presently at the weakest point it has been in 440 million years. That's how we can 'see' why the Earth is presently locked into an Ice Age epoch, and has been so for a few million years, with the glaciation getting evermore severe. The ice ages began roughly 2 million years ago, on the long galactic down slope of the last 100 million years.

Progressively more severe

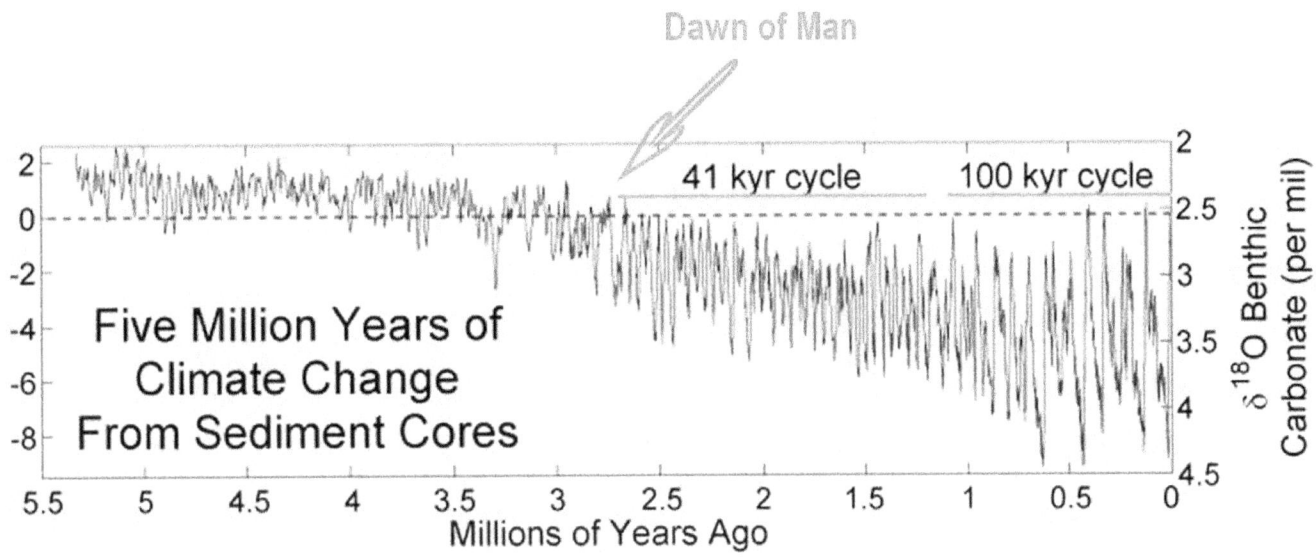

And as I said, the ice ages have become progressively more severe over time.

During the last half-a-million years

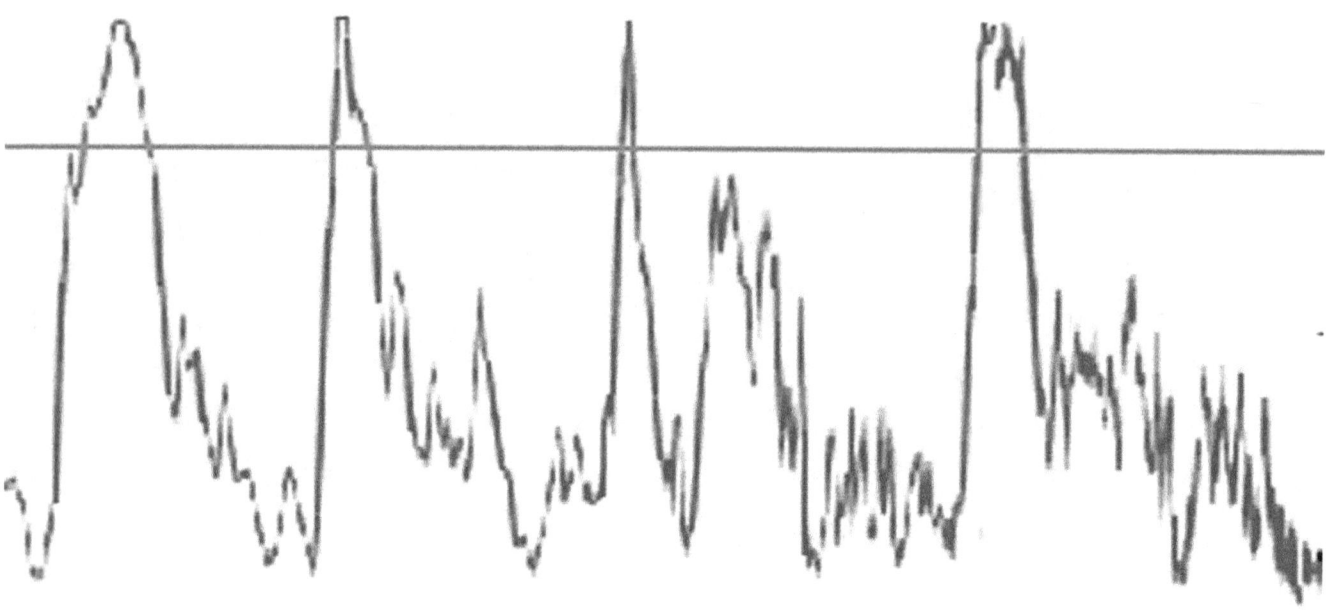

The result was that during the last half-a-million years there was so little plasma density available for the Sun, that its plasma-concentrating magnetic fields could only form during the peak of the interstellar resonance in the plasma streams that connect up our solar system with our solar system..

Only 15% of the interstellar resonance cycle

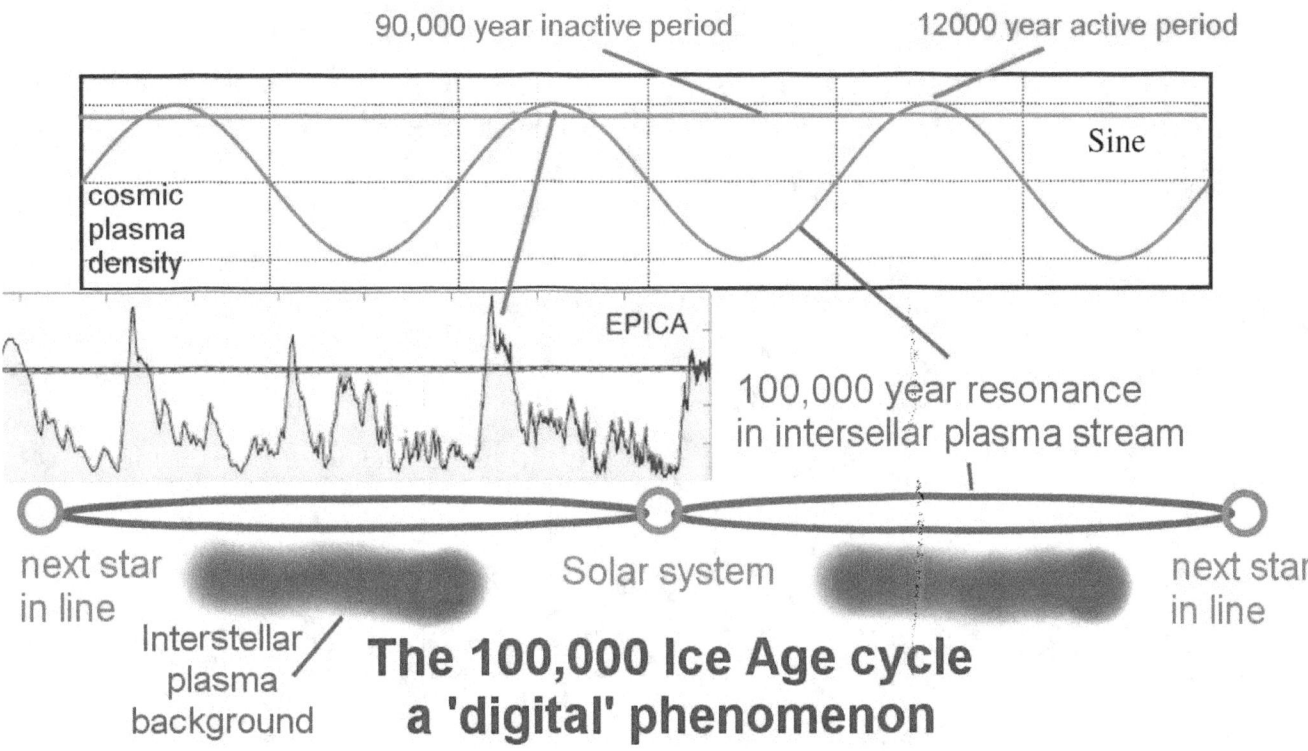

We know from ice core records from Antarctica that the peak times, during which the magnetic fields form that boost the Sun to its currently high-intensity state, cover only 15% of the interstellar resonance cycle, so that for the remaining 85% of the roughly 100,000-year resonance cycle for our solar system, the magnetic fields cannot form for the lack of sufficient plasma density in the system.

During those weak times

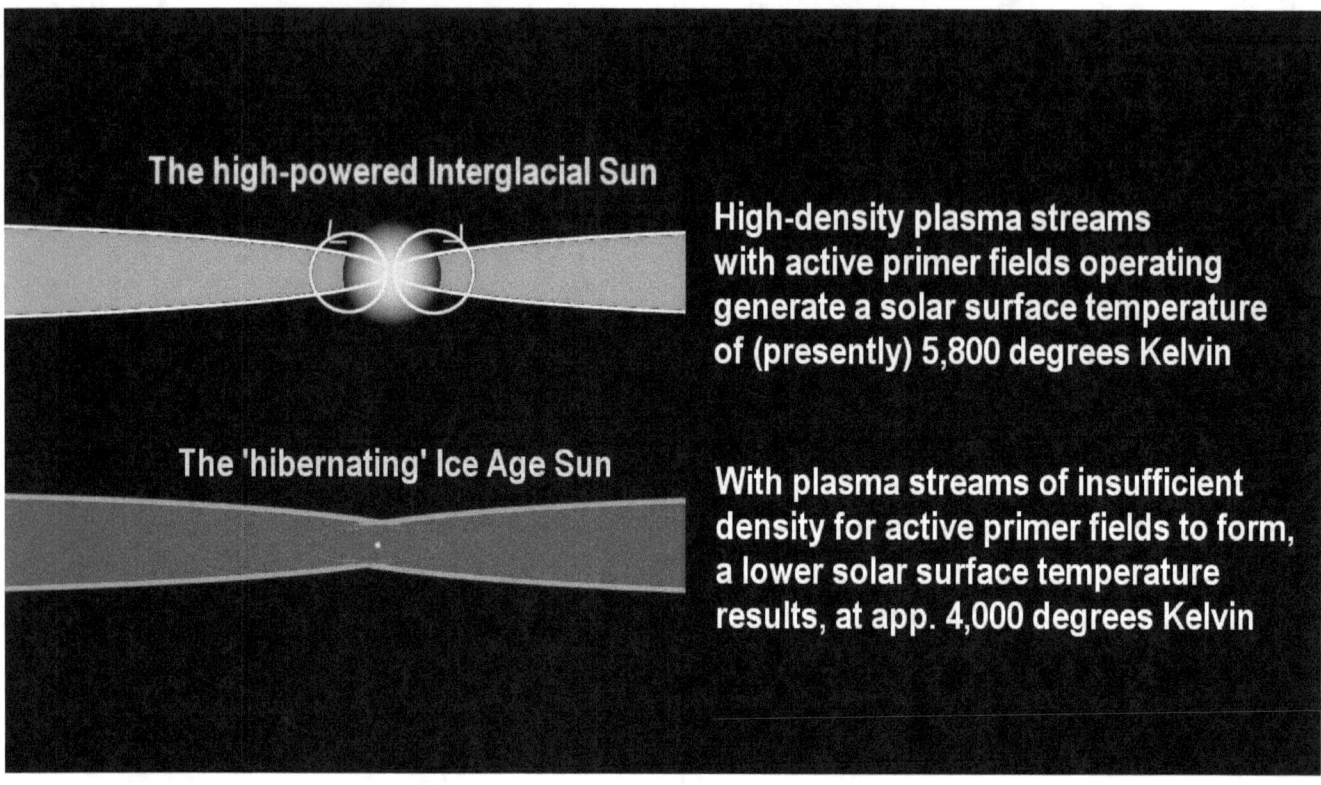

During those weak times when the primer fields do not form, the plasma flows less focused. It loosely flows around the Sun. Without dense plasma around it, the Sun goes into 'hibernation.' The hibernation causes large-scale glaciation on the Earth.

Global cooling had exceeded 40-fold

As I said before, the ice cores tell us that during the previous glaciation period, the global cooling had exceeded the cooling of the Little Ice Age period 40-fold. This is big.

Cooling reflects a 70% weaker Sun

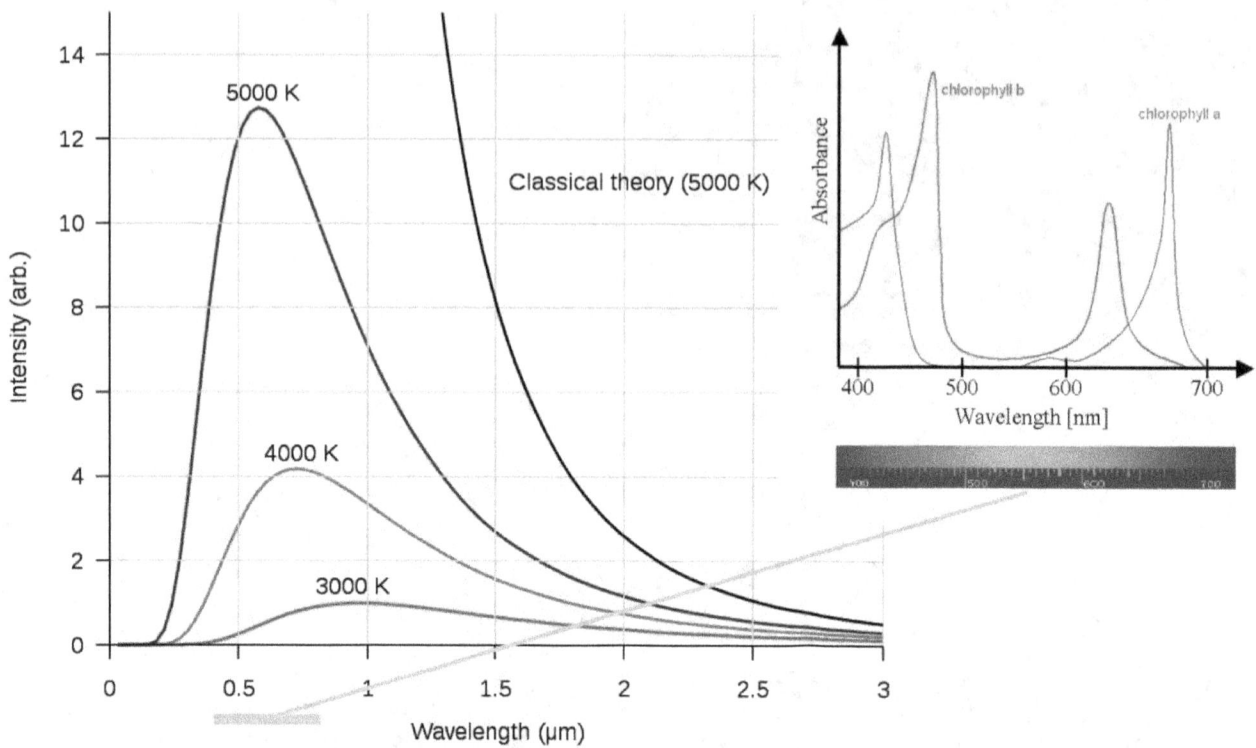

This deep cooling reflects a 70% weaker Sun, with 70% less energy being radiated. The deep reduction in solar energy being radiated corresponds to a temperature reduction on the surface of the Sun from the present 5,800 degrees Kelvin to about 4,000 degrees.

The vast majority of the stars

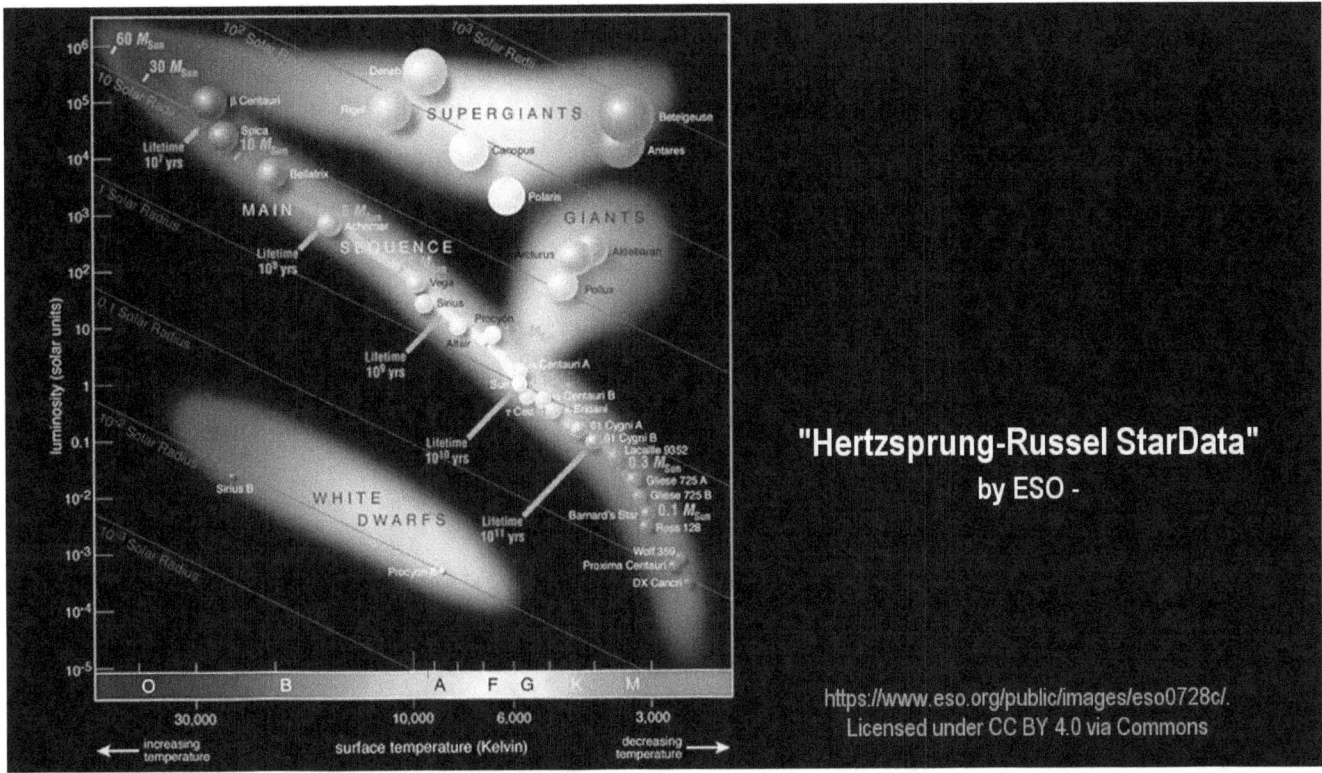

Stellar mapping tells us that the vast majority of the stars in the galaxy operate in the 4,000 degrees range and slightly below it, which appears to be a kind of default level in our galaxy at the present time, as the majority of the stars in the galaxy fall into this category. Our Sun is located slightly above the edge of this category.

Our Sun, as a rather mediocre star

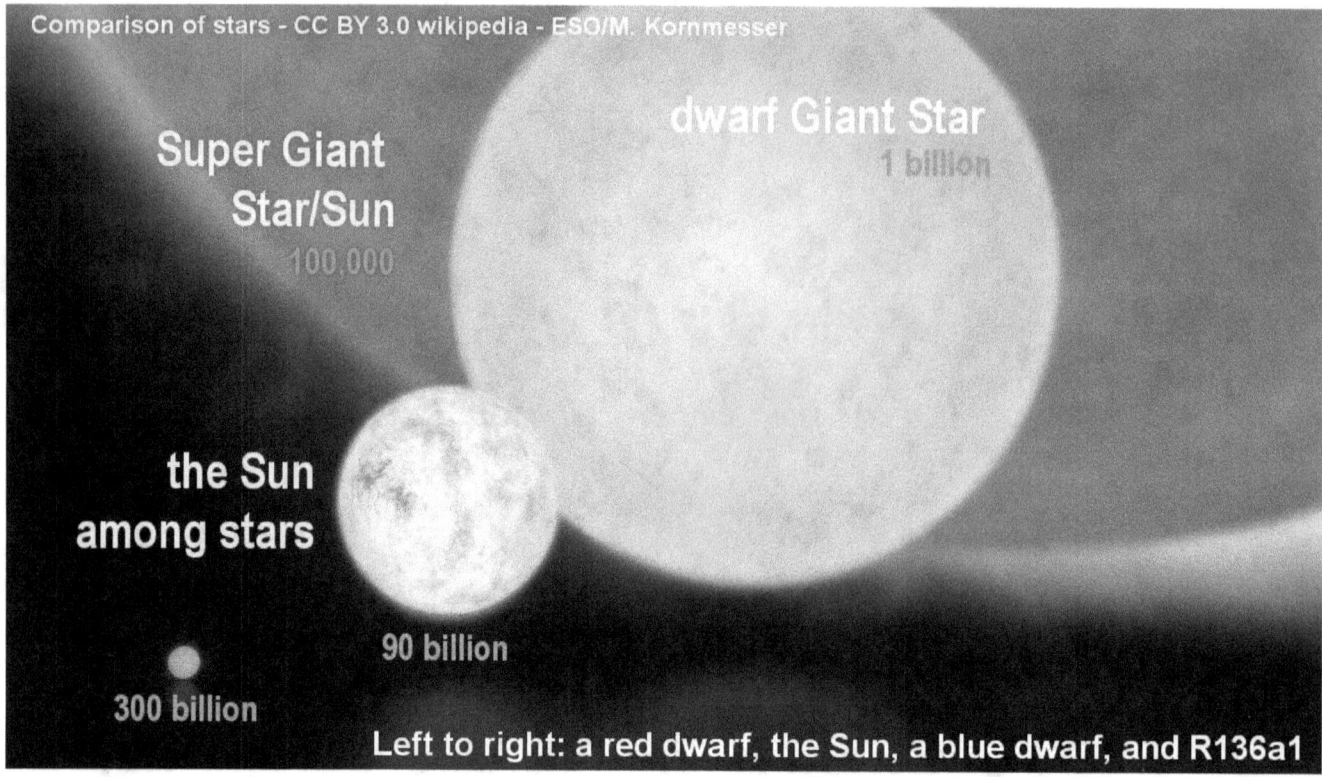

It is not unreasonable to assume that our Sun, as a rather mediocre star that falls into the low end of the brighter category, can fall out this category into the lower, default category. We appear to be near the transition point right now, for this fallout of the Sun, into the lower category, to happen.

➢ Seen the interglacial climate diminishing

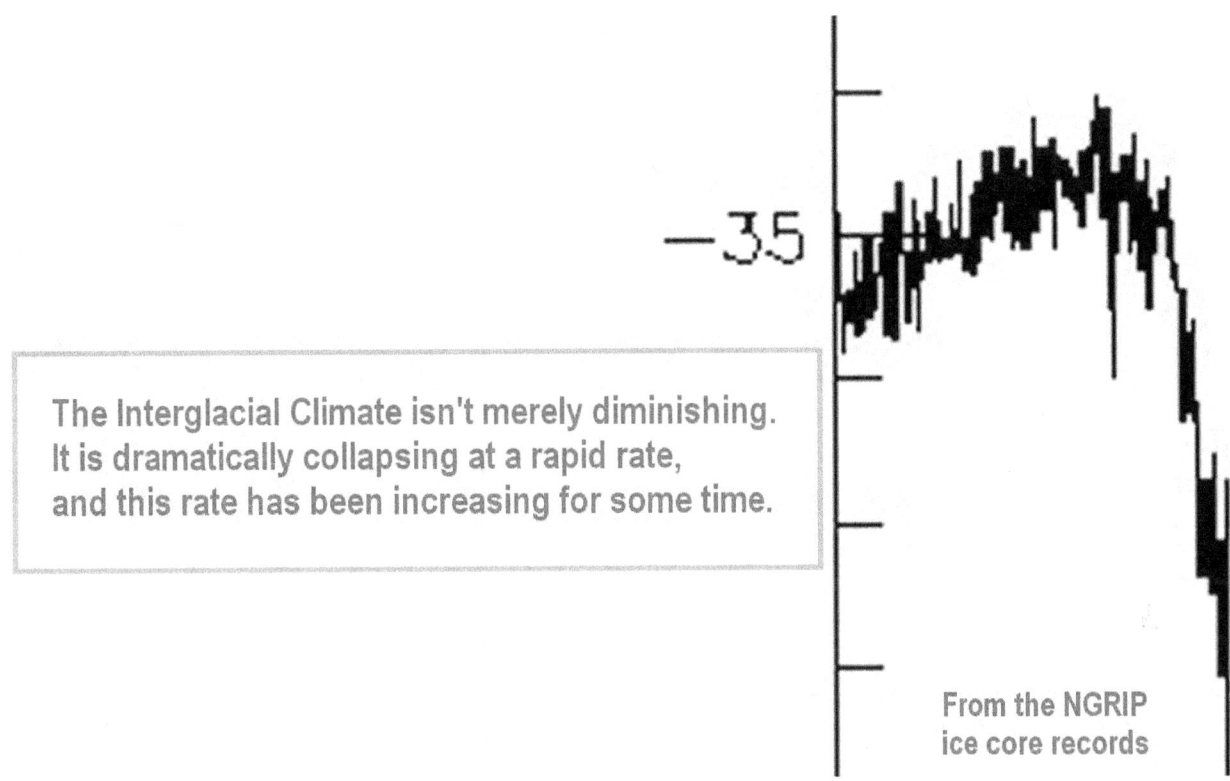

The Interglacial Climate isn't merely diminishing.
It is dramatically collapsing at a rapid rate,
and this rate has been increasing for some time.

From the NGRIP ice core records

We have seen the interglacial climate diminishing significantly already, from the interglacial optimum onward.

By everything that we see happening in escalating climate anomalies, we appear to be in the final phase towards the turn-off point.

And that's where the problem for humanity begins.

No one in general society is aware

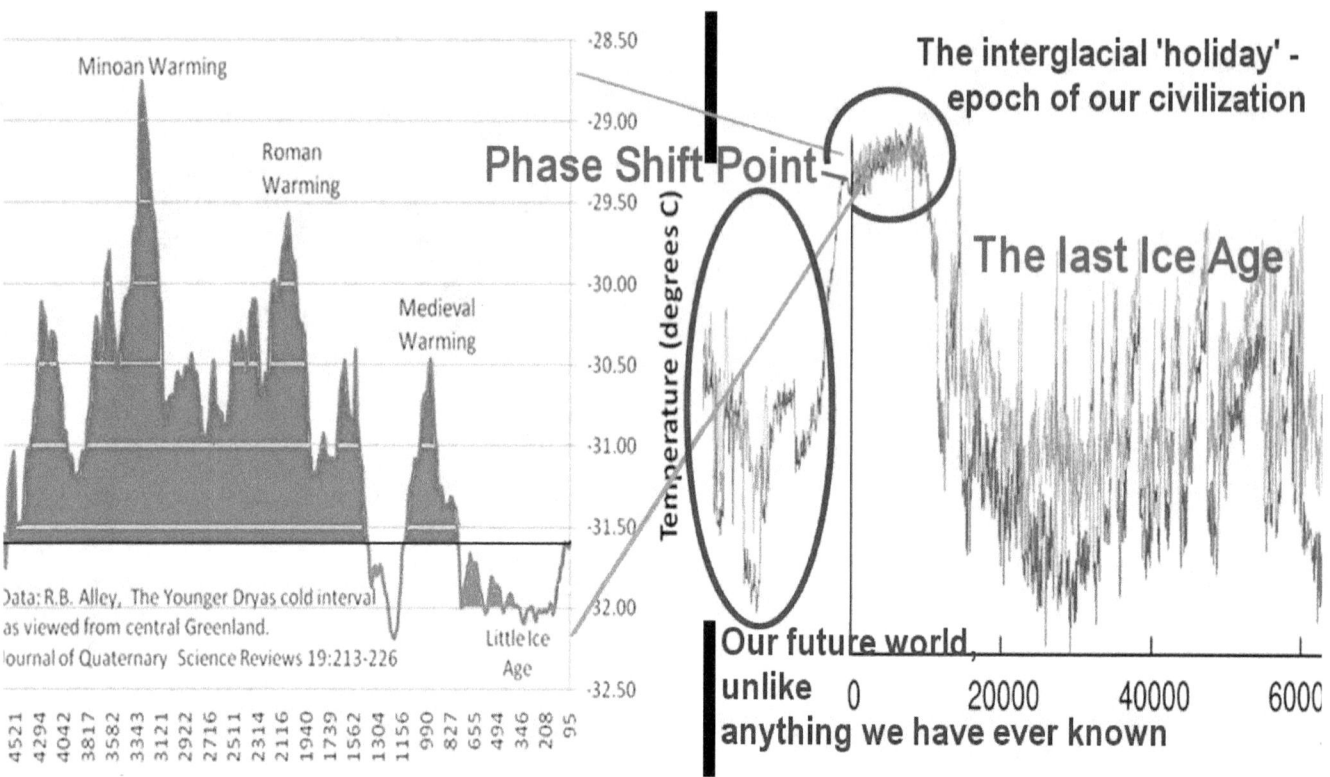

No one in general society is even faintly aware of the vast scope of the Ice Age consequences. The big climate anomalies that now affect evermore countries and regions with evermore cold, snow, flooding, storms, and droughts, appear on the surface as forebodings of another Little Ice Age coming up. In real terms they are merely escalating fringe effects that are too small to stand out on the larger scene.

We still live in an interglacial climate with a few ripple effects happening within the interglacial environment that are only significant in relationship to the interglacial climate that we have been taught to regard as 'eternal,' rather than fragile; that we cling to as absolute, globally as a society, even while our precarious interglacial climate is already collapsing.

When the phase shift happens that takes us out of the interglacial environment, everything that we have experienced, even the worst anomaly, no longer applies. We will then live in a different world with a 70% weaker Sun, and 80% less precipitation. The new environment spells mass-death for humanity by starvation, without fail, because no one can live without food, and food cannot be grown under glacial conditions when much of the Earth becomes largely an Ice Planet.

If we have not built us a new world by the time the phase shift happens, with technological infrastructures that the Ice Age cannot touch, the whole of humanity might become extinct, as it nearly became once before, and potentially many times before that.

We are the 8th human species

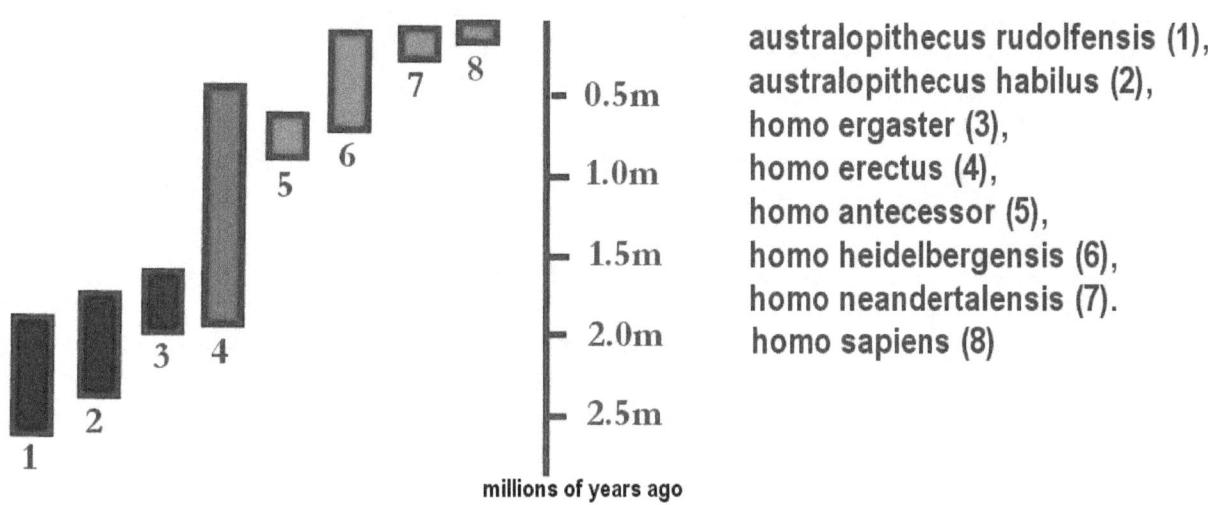

We, the homo sapiens (8), are the only surviving, and the shortest lived of all the the human species, at barely 200,000 years of age.

We are the 8th human species and the only survivor through the long string of Ice Ages in our past. We are also the only species of humanity that developed itself into a 7 billion world population, which now faces the task for the first time in human history, to maintain itself at the current level, technologically, through the next Ice Age that in previous times had enabled only a few million to live, existing in isolated pockets where 'natural' living remained possible.

The most critical juncture in our entire history

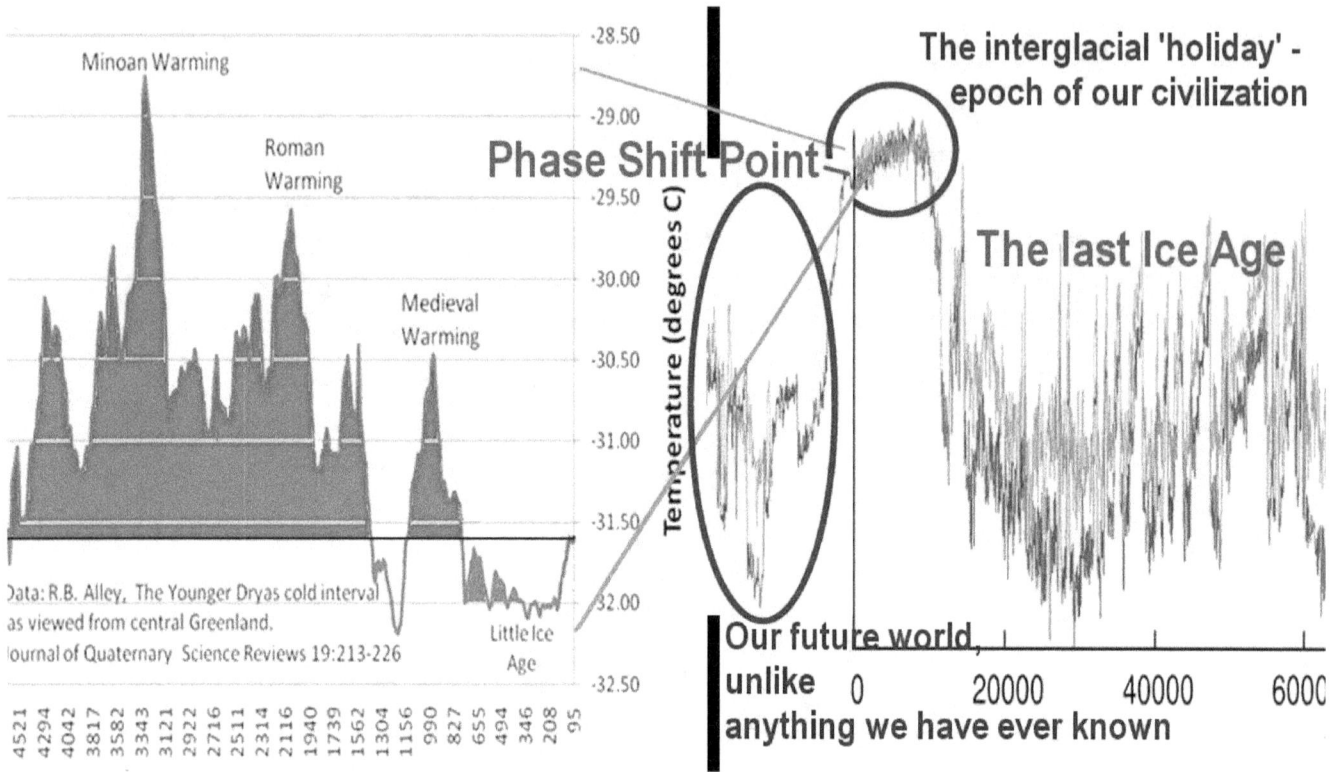

Ironically, it is here, at the most critical juncture in our entire history, that we, as the most advanced species that ever existed on this planet, have put ourselves to sleep mentally, riding the fast train to oblivion by not responding to the Ice Age Challenge, as if it didn't exist.

We have 'seen' in Beryllium records

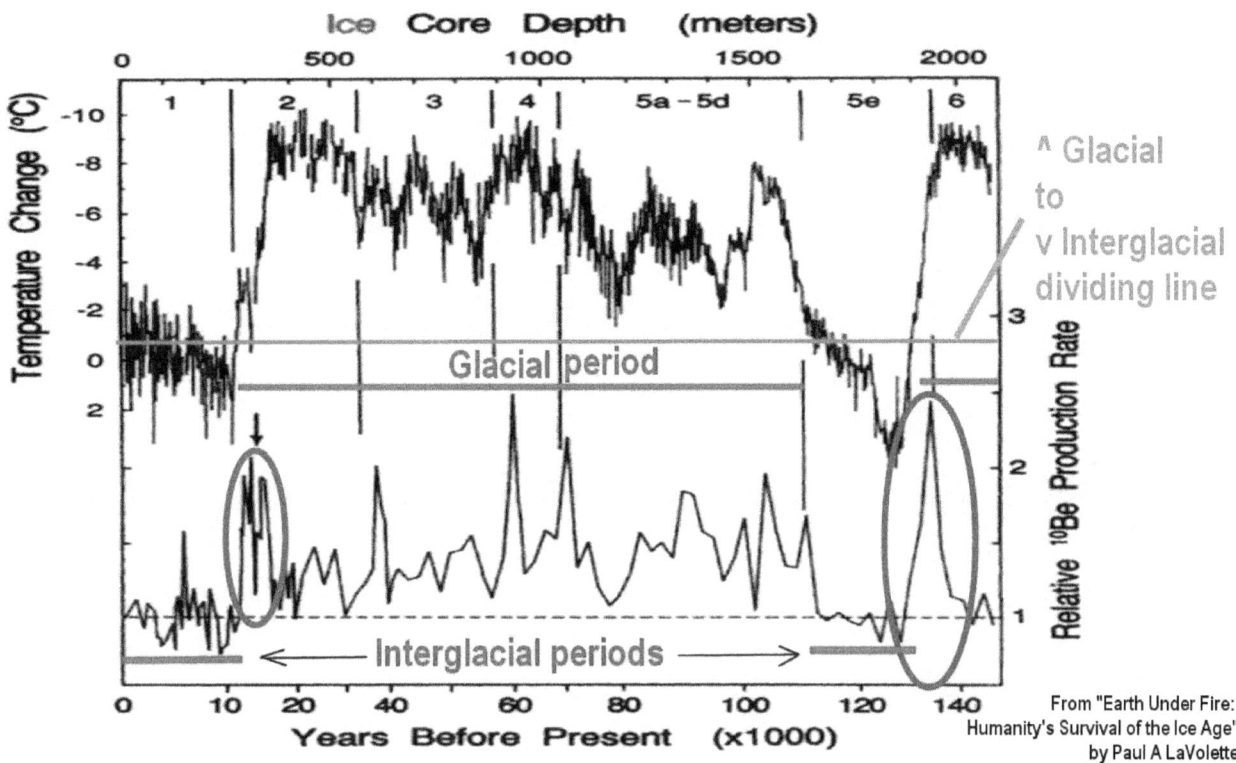

We have 'seen' in Beryllium records that the climate on Earth, including the Ice Ages, are caused by the Sun.

We have seen the Sun getting weaker

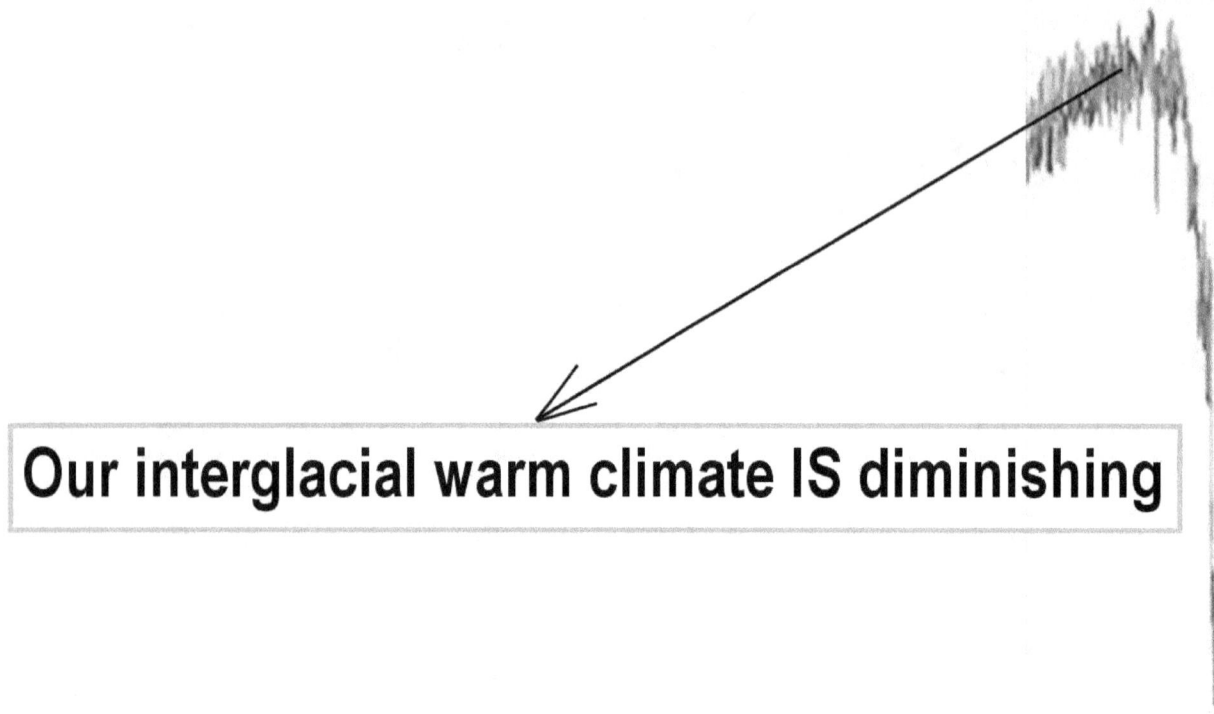

We have seen the Sun getting weaker over the last 5000 years of the interglacial climate.

Solar warming events diminishing

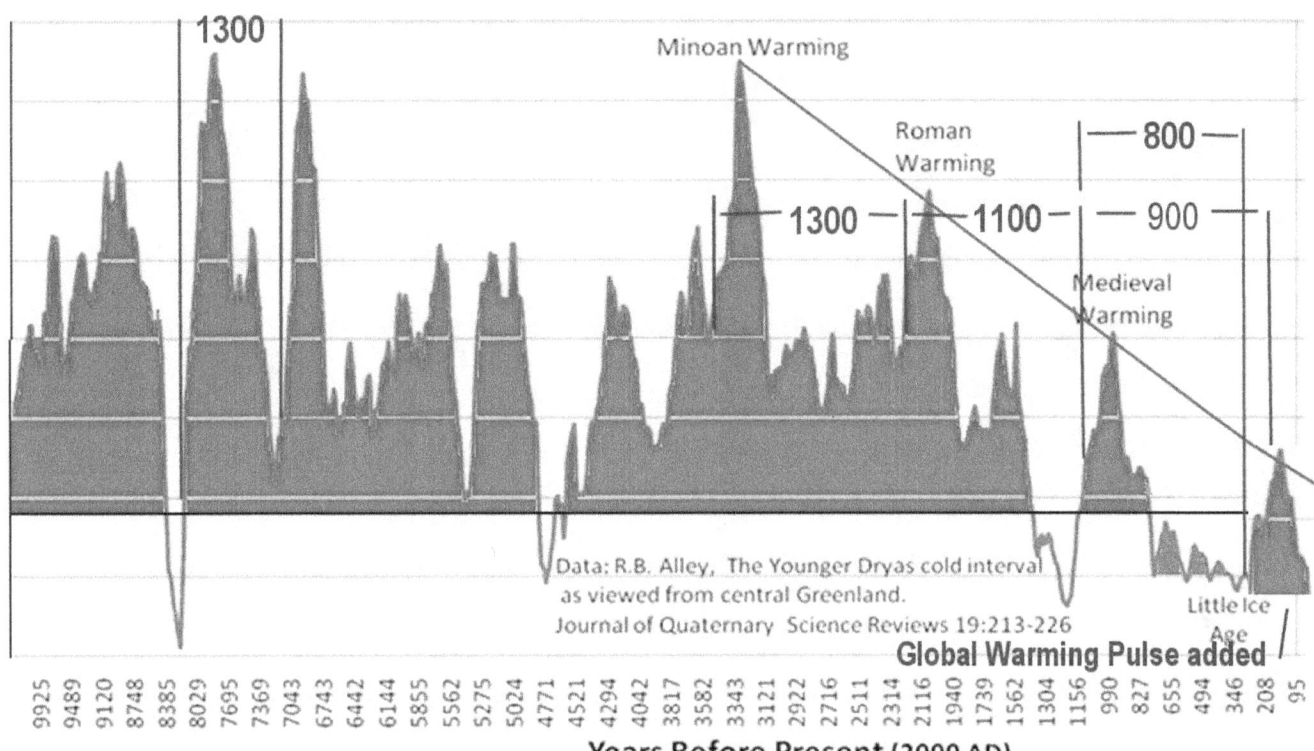

We have seen the solar warming events diminishing rapidly.

Solar minimum events diminishing

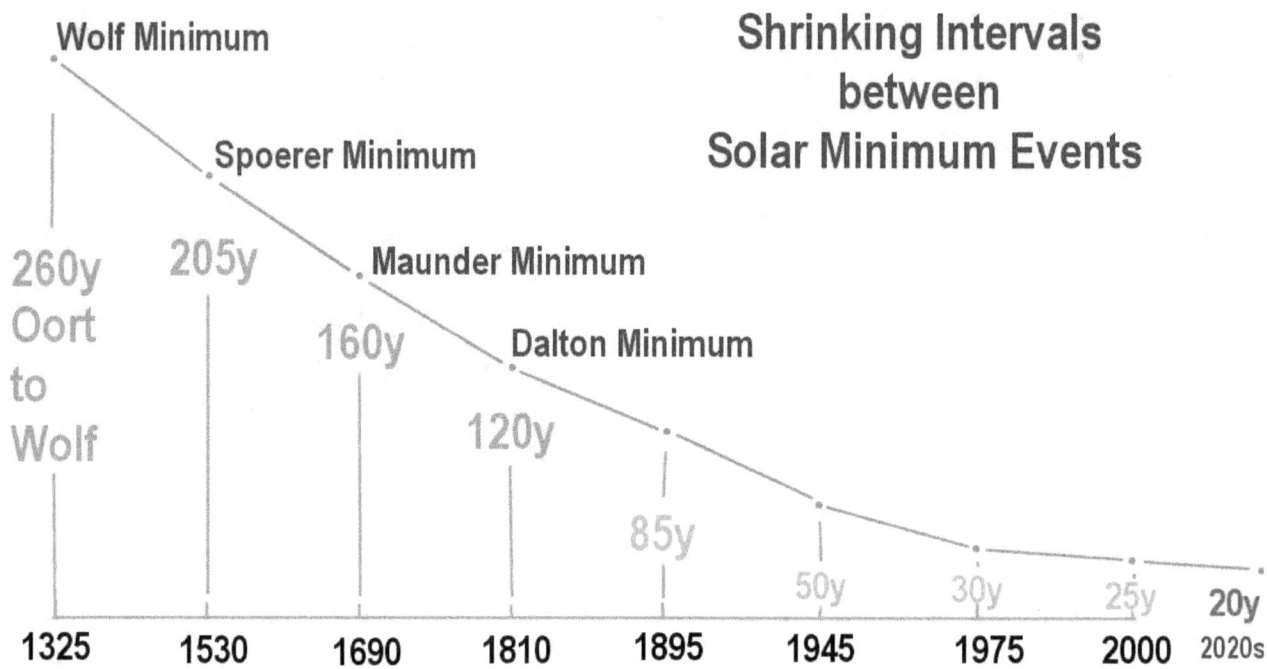

And we have seen the solar minimum events diminishing almost geometrically.

We now see the solar cycles diminishing

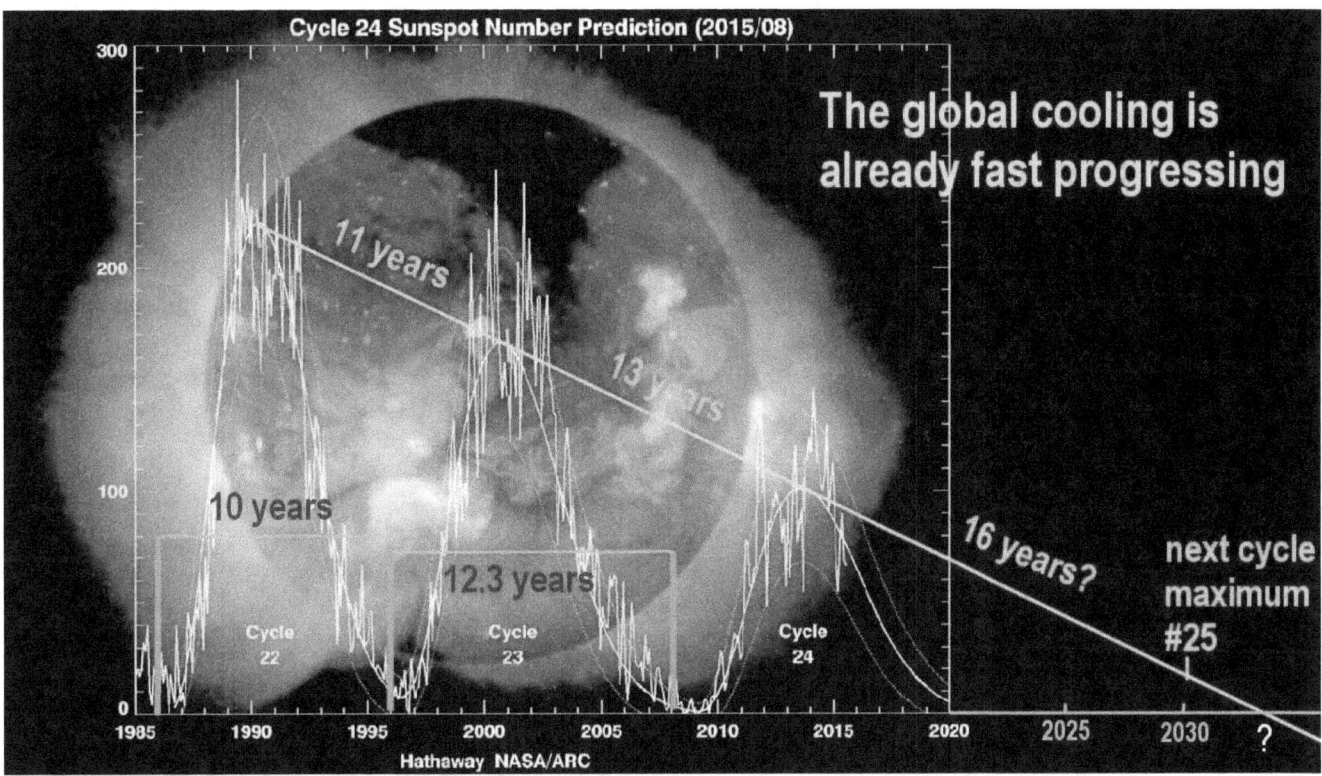

In addition we now see the solar cycles diminishing too, and their 'heart beat' slowing down.

We even see the solar-wind pressure collapsing

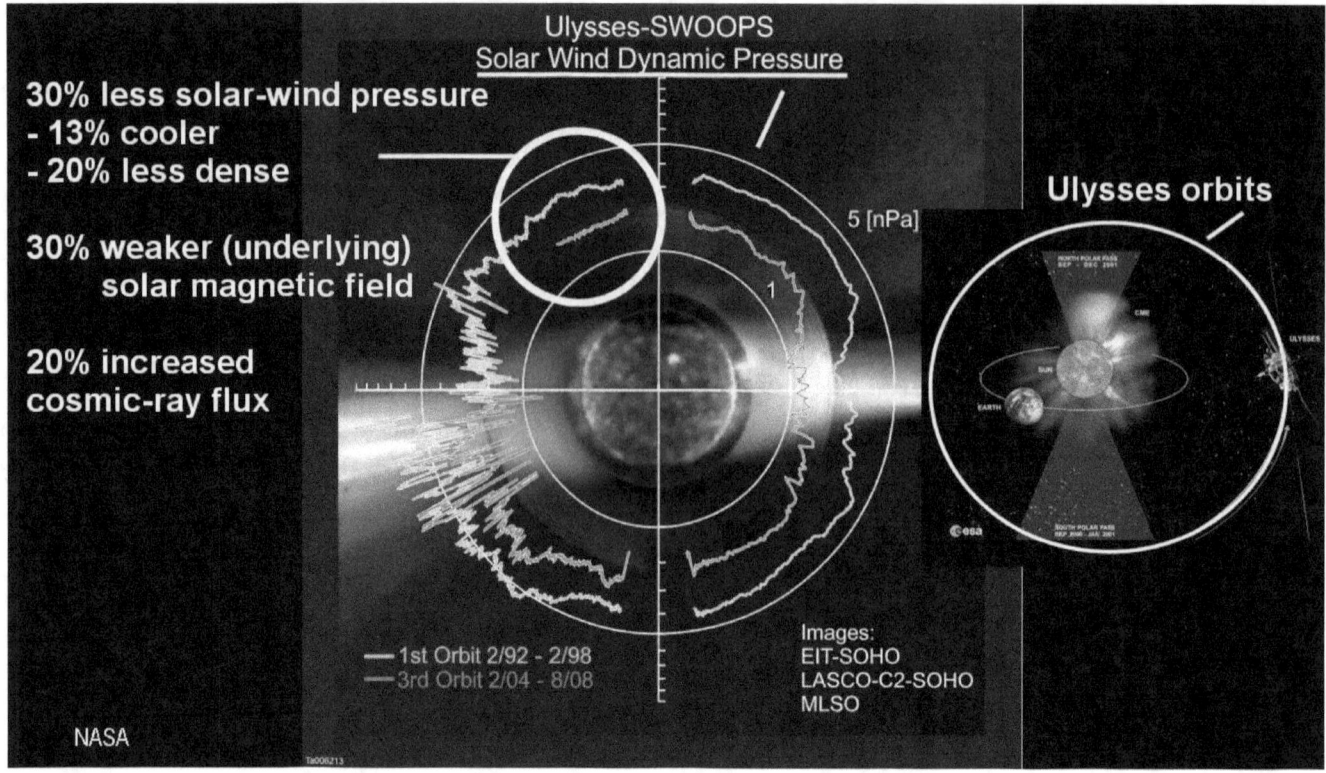

We even see the solar-wind pressure collapsing now, at the astonishing rate of 30% per solar cycle, and likewise the solar magnetic field.

We now see the solar weakening progressing

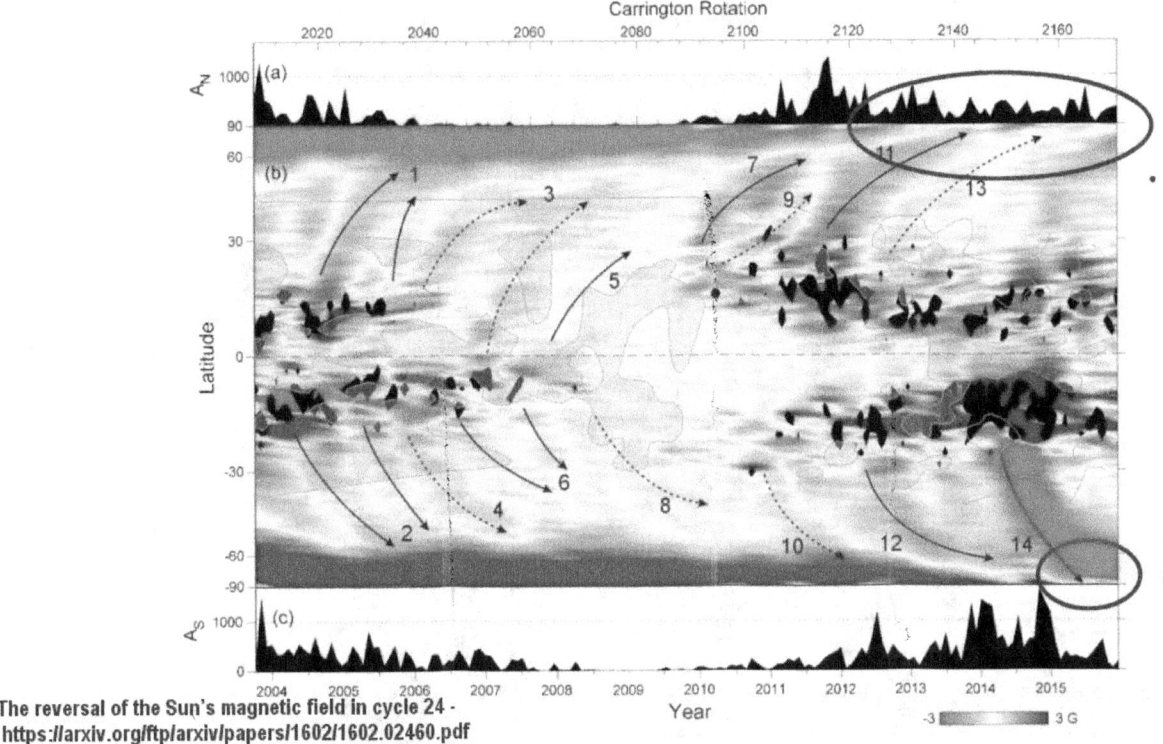

The reversal of the Sun's magnetic field in cycle 24 -
https://arxiv.org/ftp/arxiv/papers/1602/1602.02460.pdf

We now see the solar weakening progressing so fast that from 2013 onward, the Sun lost its northern polar magnetic field completely.

We have watched all of these events unfolding

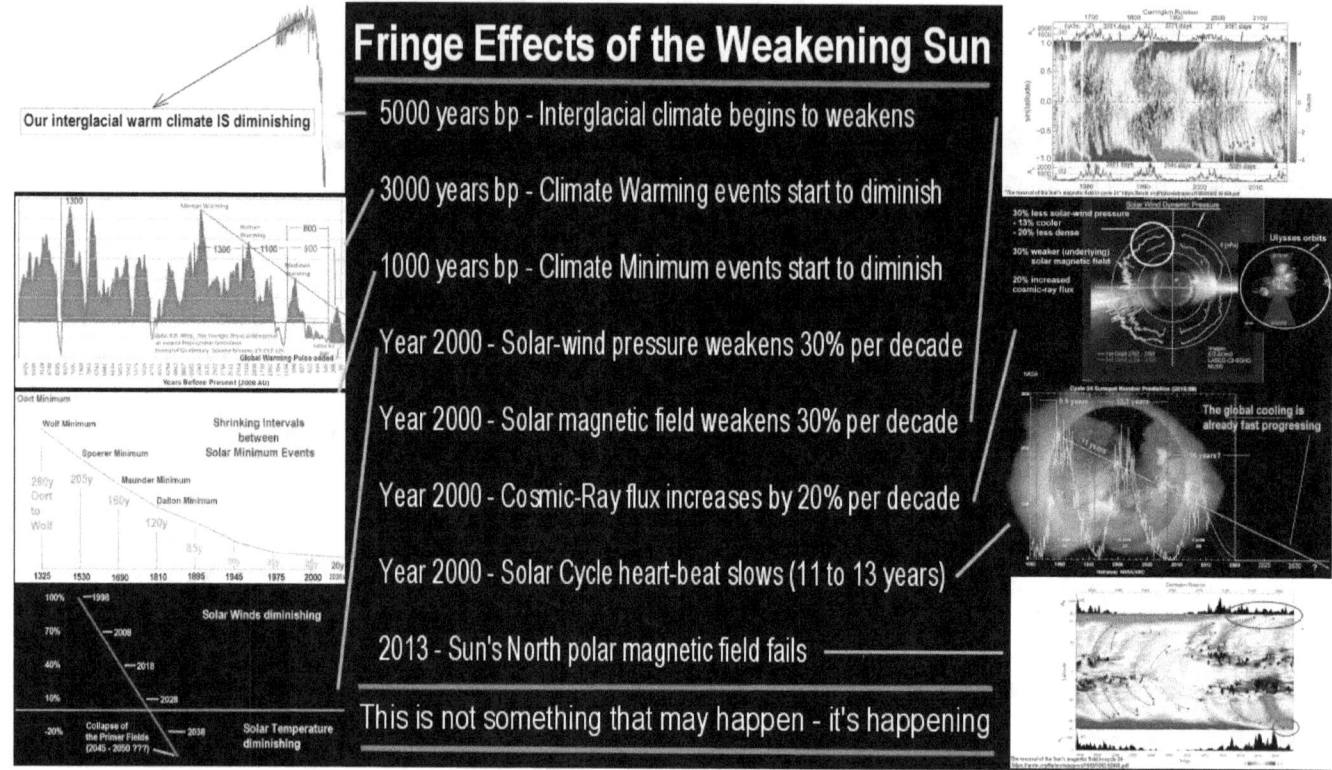

We have watched all of these events unfolding. We know that they are real, because we have measured them all, and still we close our eyes to them as if none of that had happened. We understand the dynamics of what we have measured, and their obvious consequences.

Project the consequences into the future

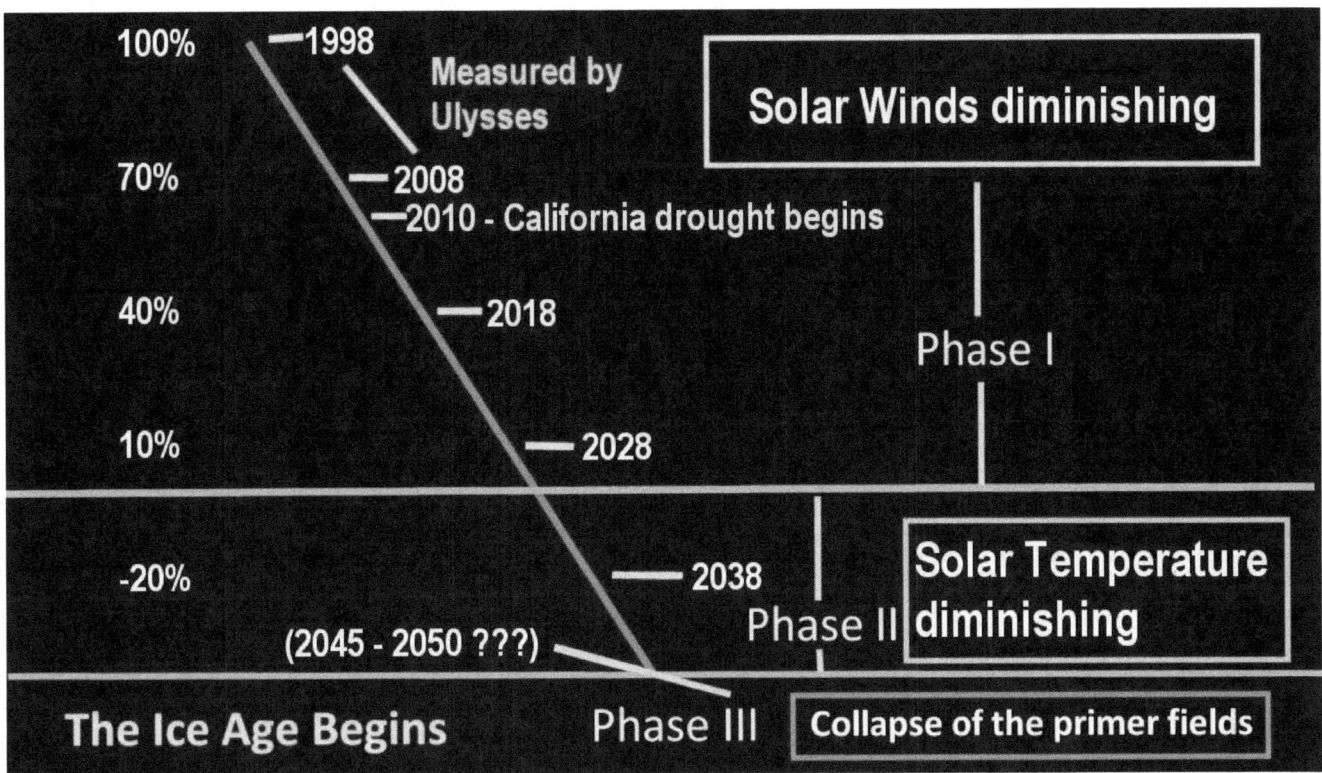

And we can project the consequences into the future to determine the potential phase shift timing, which may be as close as the 2050s, and still we staunchly refuse to stir our stumps to get the spate into the ground to assure our future existence.

We have made enormous scientific efforts

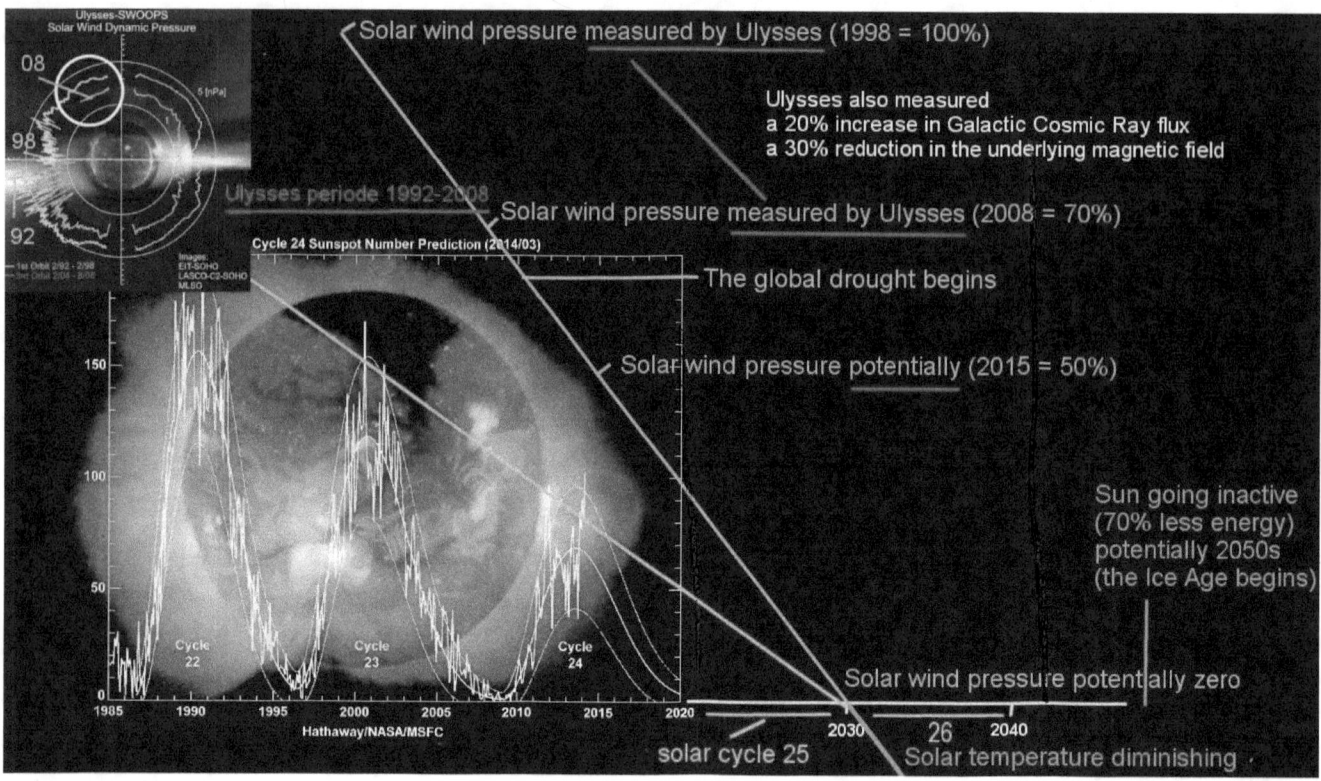

We have made enormous scientific efforts to collect the data that enables us determine the phase-shift timing into the future.

As near as 30 years from now

We also know from the acquired data that this future that is potentially as near as 30 years from now, will unfold on an uninhabitable planet that we cannot survive unless we built us the technological infrastructures that enable our continued existence.

The near future is a death-trap without

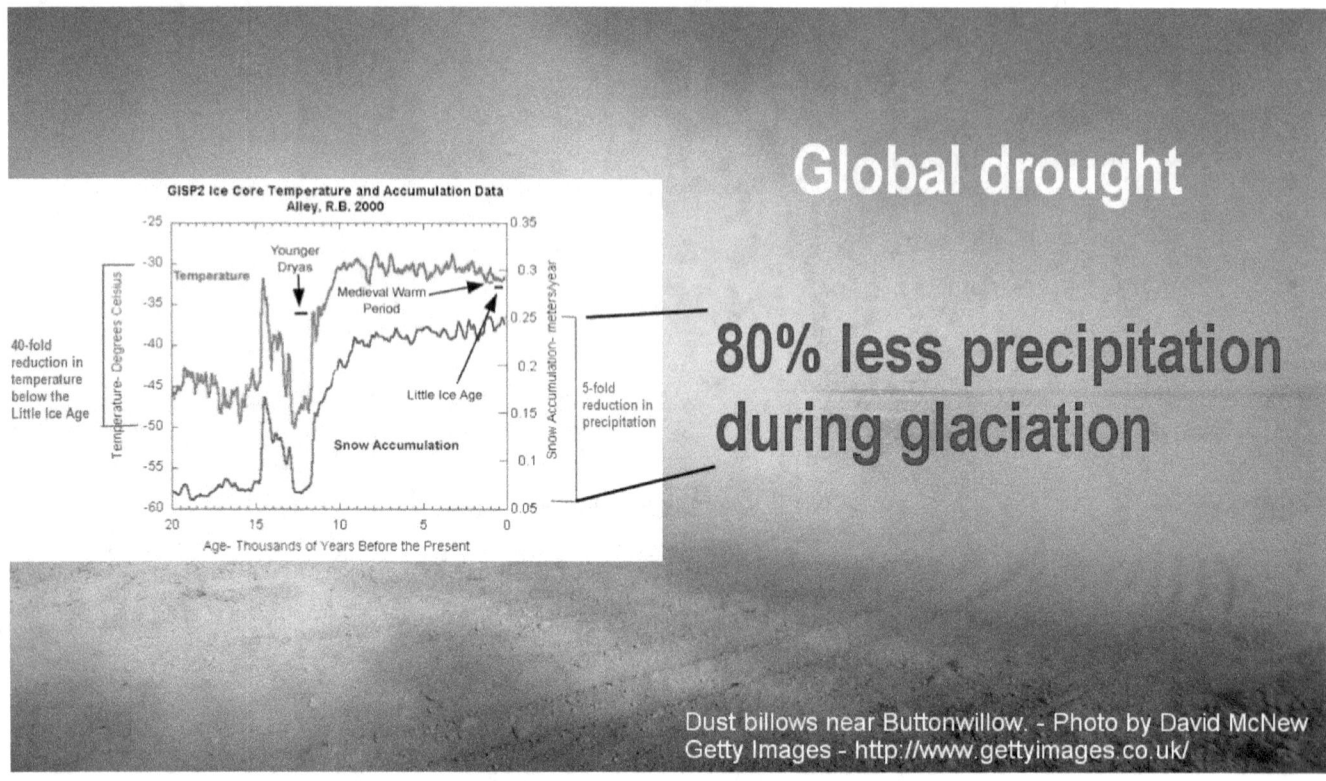

We know from the acquired data that the near future is a death-trap without the technological infrastructures for continuing human living, because the natural landscape under glaciation conditions yields nothing - as a world without food.

We remain stuck in our beloved easy chair

But even as we know all this, we remain stuck in our beloved easy chair, with an iron-fist-type commitment to remain stuck, do nothing, with the very thought banished that would built us a new world in which continued human living is possible, as if human life isn't worth the needed effort to maintain it, and our promise to our children of a bright future is as empty as we ourselves have become as a human society.

That's where we stand today. We have trapped ourselves into an empty small-minded world of utter impotence, content to be living with our eyes and ears closed, and our humanity banished, lest it would wake us up to embrace the world and humanity as a precious gem.

➤ Getting out of this trap

Getting out of this trap of many layers.
A prominent layer is false science!

Getting out of this trap

which is a trap of many layers. A prominent layer is false science!

The Sun as its own master

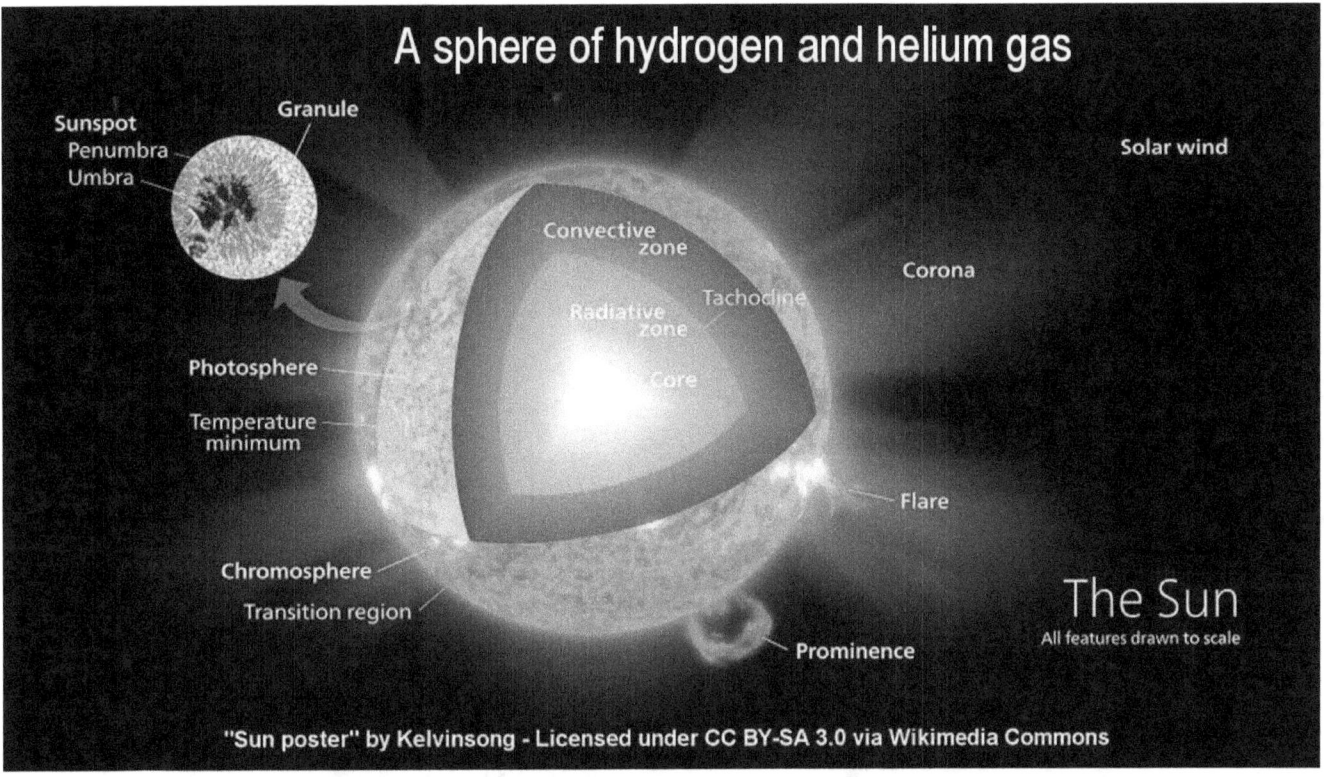

We've been taught to regard the Sun as its own master, - as an unchanging heat-engine - while the opposite is the case. In real terms the solar model that humanity is taught to hail like a God, is so full of holes that it puts the makers of Swiss Cheese to shame.

But why should we remain stuck in this imprisoning box created a hundred years ago by small-minded men who were trapped in defending a dying political system?

Let the magic tales go

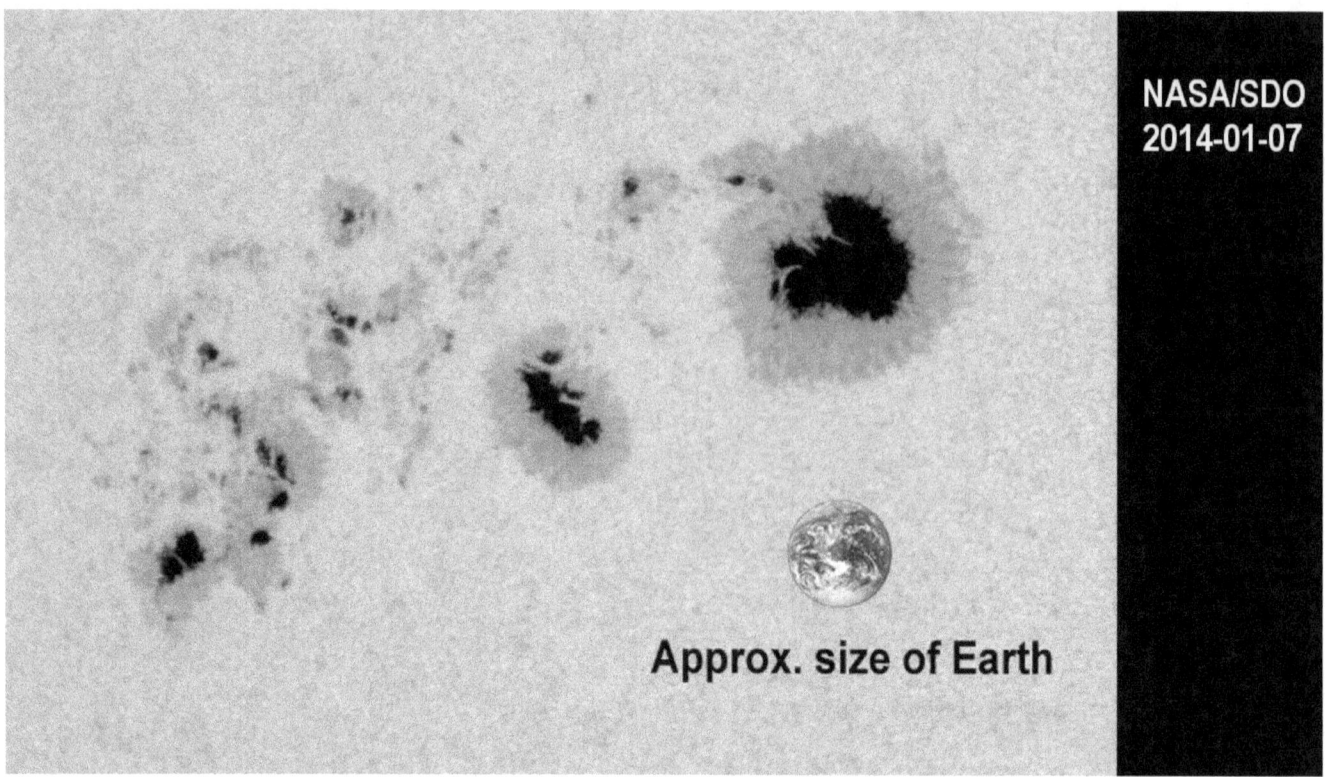

Society has grown up too far in those hundred years, as to cling to such archaic science tales like the fabled dark energy that blocks light, whereby, so it is said, the sunspots appear dark at the umbra.

Let the magic tales go. Close the book of magic.

Magic tale of the universe exploded

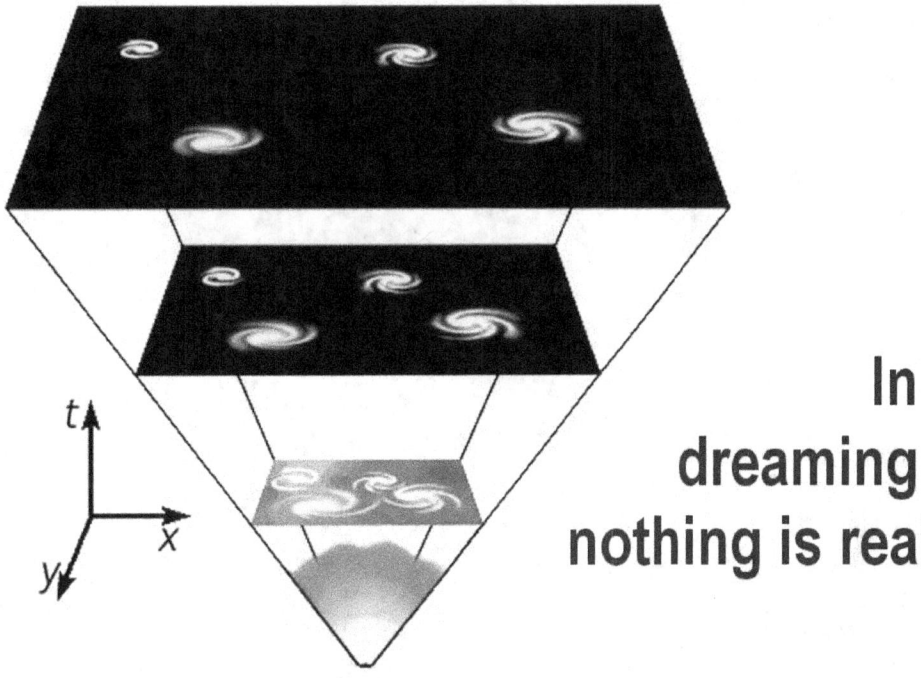

Also close the book on the famous magic tale of the universe having exploded out of nothing in the span of a millisecond. The real universe is much more understandable than that.

The real universe is a plasma universe

The center of the Milky Way, at the center of the Big Bang explosion of the universe

The real universe is a plasma universe that is universally self-creating and self-developing, which synthesizes its own atomic elements, and does so constantly. The Big-Bang magic isn't needed.

The Big-Bang red-shift a deception

The Big-Bang foundation, the measured red-shift that supposedly proves that the universe is exploding, is a deception, since the red shift phenomenon is merely the result of energy depletion in photons on their path across millions of light years of space and across large fields of plasma on the way.

A new world comes into view

The center of the Milky Way, at the center of the Big Bang explosion of the universe

When we break out from the frame of magic in science, to what is actually real, a new world comes into view with the sobering awaking.

The Plasma-Sun is actually physically possible

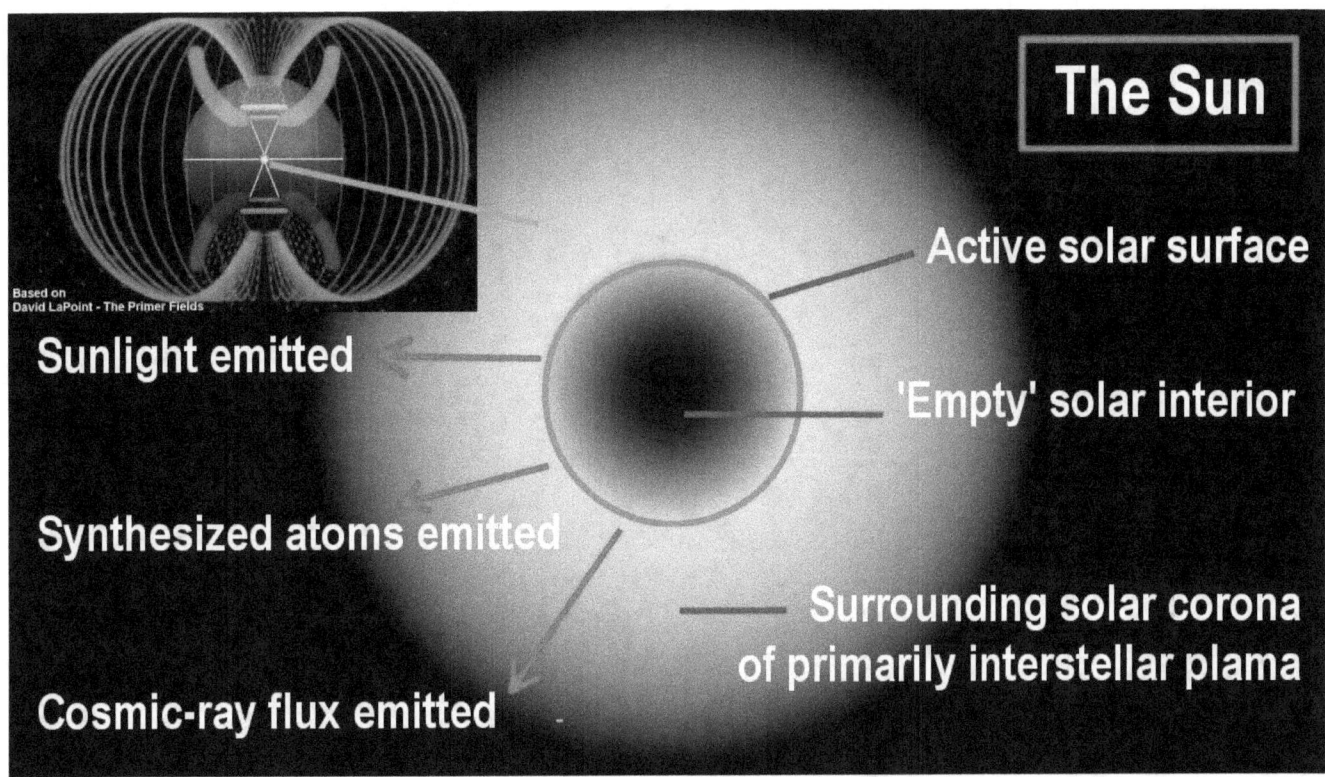

The Plasma-Sun theory is the only solar theory that is actually physically possible and supports all known solar events and features. The theory is simple. It works, and its key features have been replicated in numerous laboratory experiments.

The Gas-Fusion-Sun concept

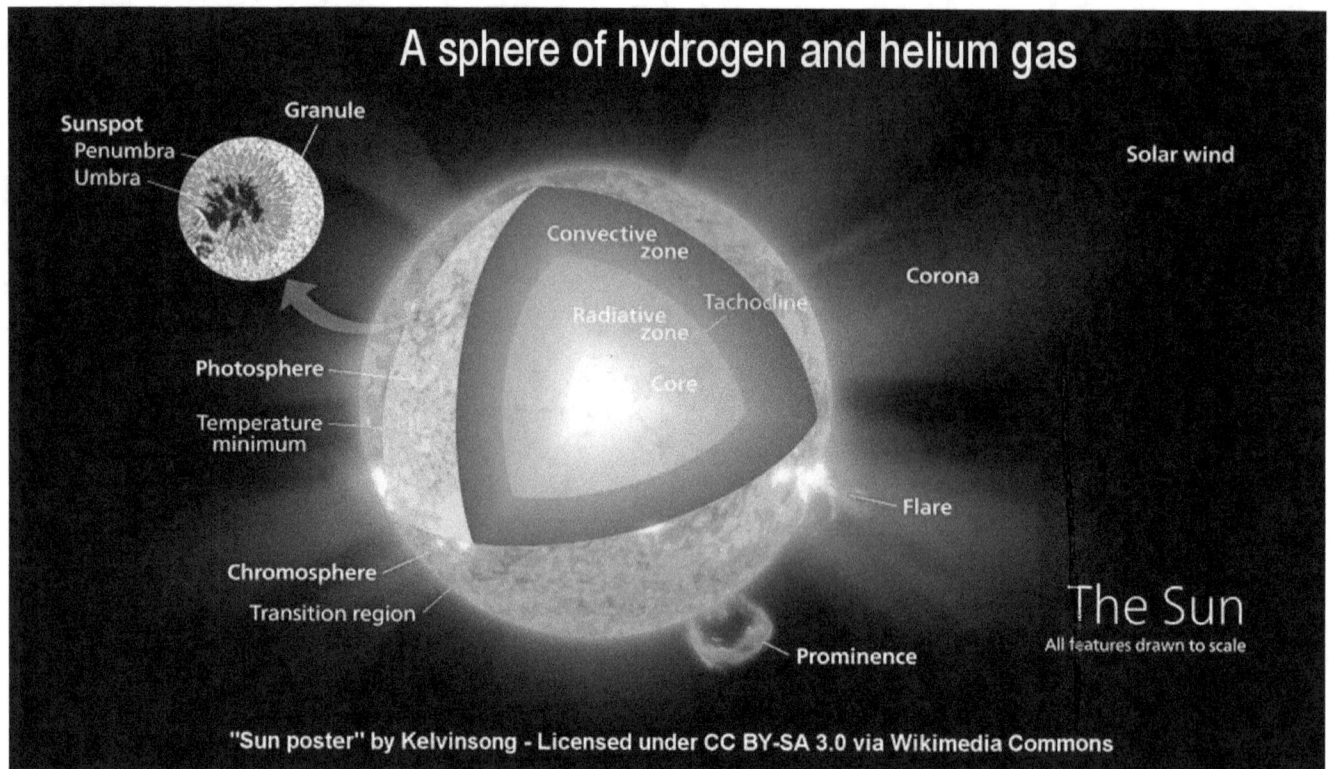

The Gas-Fusion-Sun concept, in contrast, doesn't have a workable foundation. Far from it. It is a doctrine built on magic that one is required to accept on faith, such as the electron degeneracy magic that is deemed to make a gas sun physically plausible. Even the fabled nuclear-fusion-energy production in the solar core is magic, which every laboratory experiment has proved to be not possible, and has also proved that nuclear fusion is an energy-consuming process.

The Plasma-Sun brings the solar system into the universe

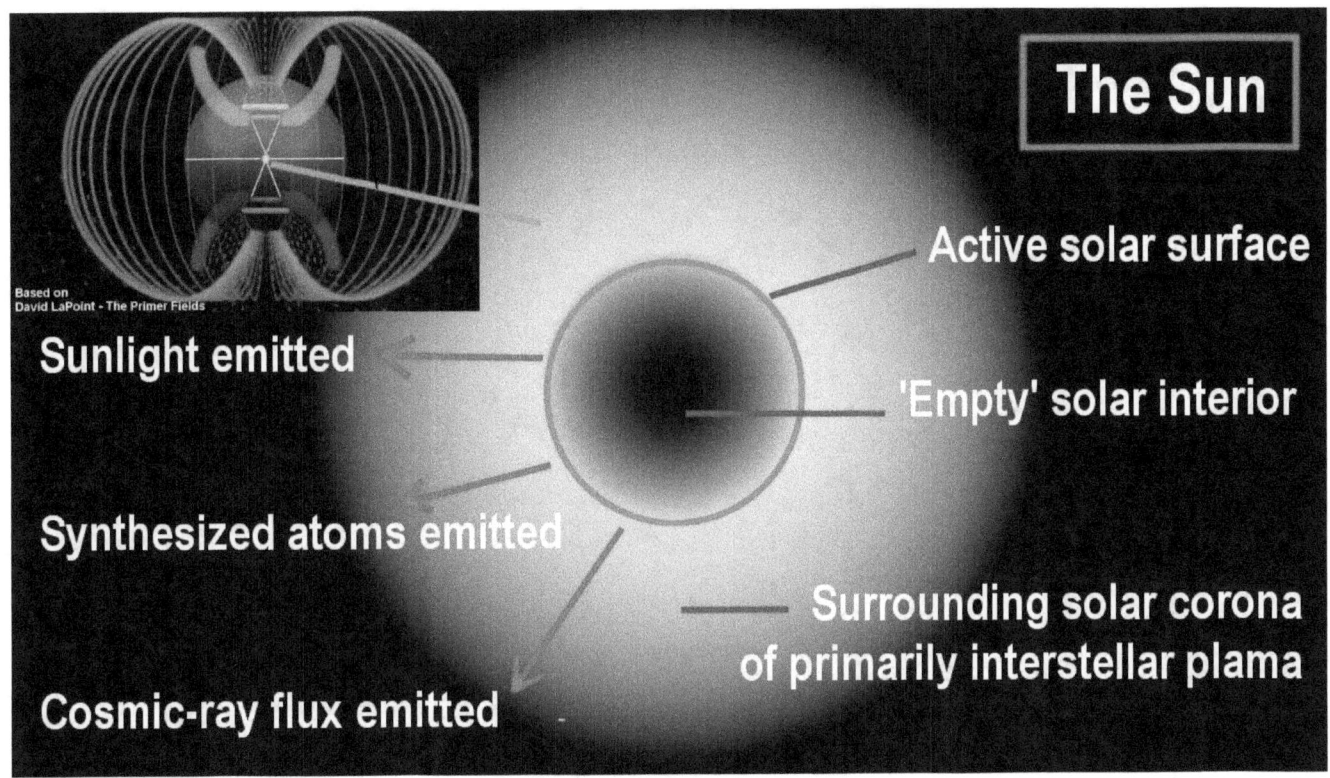

The Plasma-Sun theory doesn't have these problems. It brings the solar system into the universe as an integral element of it. No magic is required. Every known principle supports it. And most important of all, it takes the Ice Age dynamics out of the box of magic and makes it measurable and understandable and predictable.

The gas-sun is the most deadly doctrine

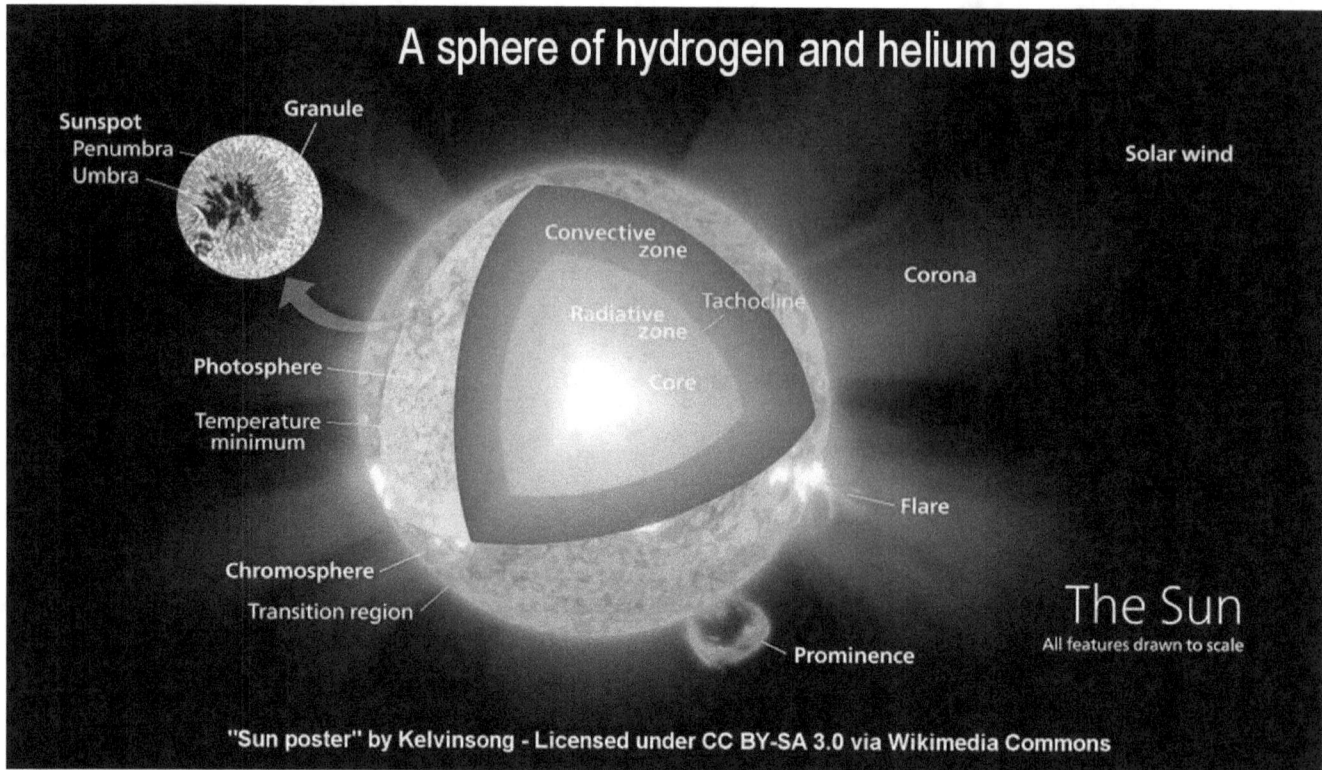

The doctrine of the gas-sun magic is the most deadly doctrine ever created. It is deadly by its doctrine of the constant Sun. The constant Sun myth blocks the Ice Age recognition. It prevents humanity from building itself the needed technological new world that the Ice Age cannot touch.

So, who is afraid of the Ice Age?

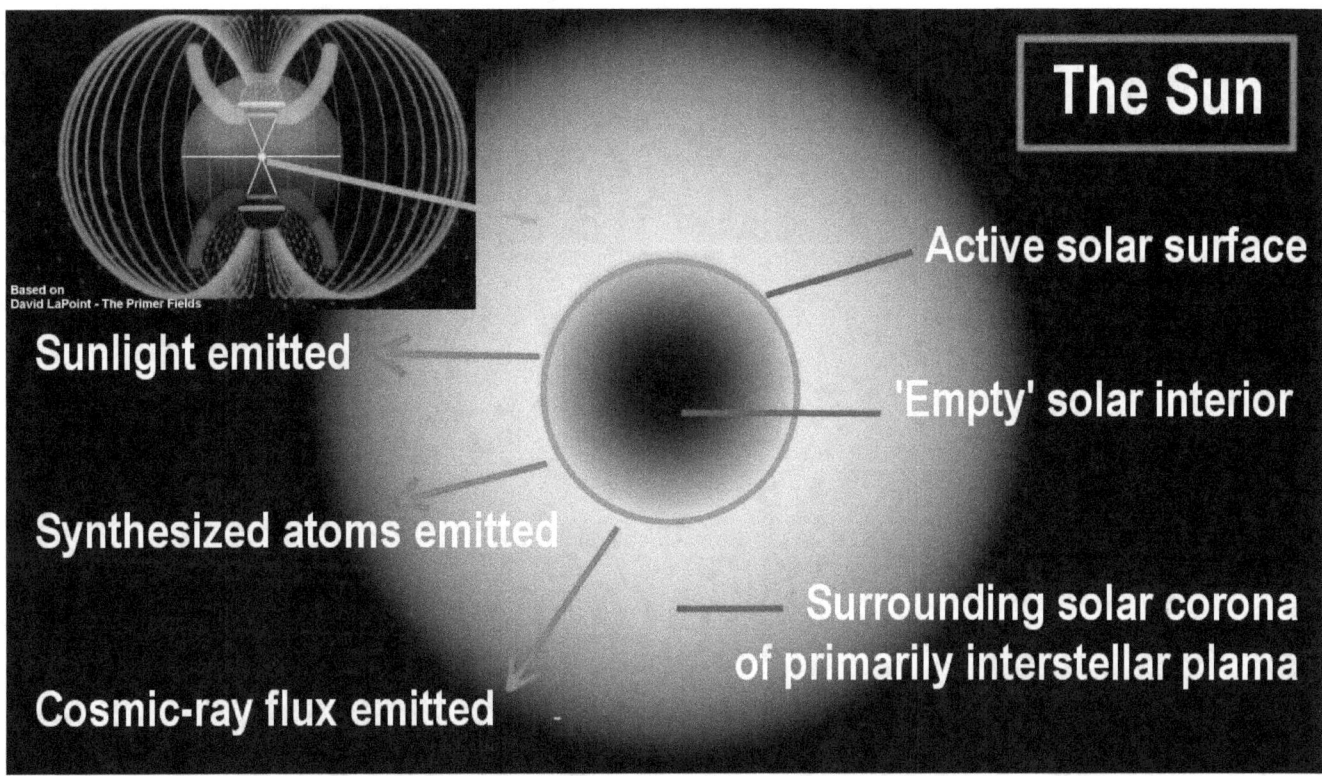

However, when we step away from the dark-magic-science, into the landscape of real science, the Ice Age is not scary at all, but comes to light as a bugle call to humanity to become alive as human beings and to experience its dormant potential in the process of creating itself a new world with greater security, grander prosperity, and with cultural advances that give the terms "renaissance" and "happiness" a new meaning, and give the very concept of a future, a golden shine.

So, who is afraid of the Ice Age? Not I.

Who wants to live as dead men walking, as humanity presently does? Not I.

Who wants to live choked with debt and rent slavery, and food prices that are reaching for the sky? Not I.

To be part of the Ice Age Renaissance world

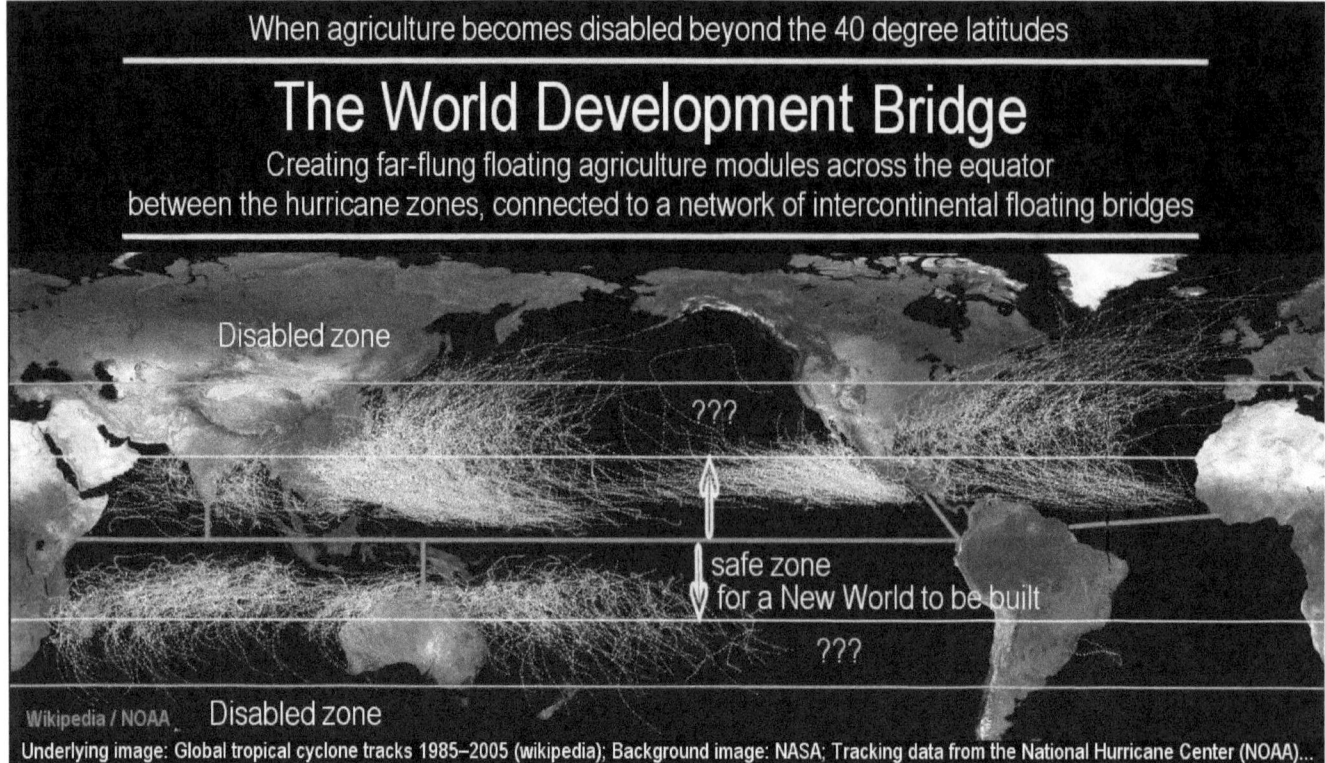

I am not afraid of human development and its resulting freedom for humanity to live evermore, with songs of joy on its lips for the achievements wrought. I would like to think that there are probably billions of people who would love to be part of the Ice Age Renaissance world that lays before us, who would gladly become involved in creating it, if the opportunity for it existed.

With this kind of future on the horizon, who in society would choose to offer their life to the war-criers to die in the ditches of a destroyed world? I like to think that those are few and far between who opt to waste their life that way, in comparison with those who opt to come to life as human being by experiencing the grandeur that the Ice Age World-Development has to offer.

Just imagine producing 6,000 new cities

Just imagine being a part of the project of producing 6,000 new cities for a million people each, on land's that do not yet exist, produced in automated industrial processes, fabricated with high-quality material that exists in infinite abundance, process ready on the ground, processed with energy technologies that have been sitting on the shelf since the 1950s, and with fuel resources that may never be depleted. We lack nothing but the love for one-another as human beings to make the grand potential that we have, a reality.

I think, this is the route that we will choose when we manage to awake us up from the dream-world of false science magic, false politics, and false ideologies, to what we are and can yet become.

The time to get started is now. We have 30 years grace to get the job done. The goal is to make the grade, and to earn us a Double A-plus on the report card when the Ice Age exam is called 30 years from now. Our success will determine whether we will live and have a future, or not. Much hangs in the balance here.

➤ **The power of the universe is aiding us**

| The power of the universe is aiding us |

The power of the universe is aiding us

Increase in cosmic-ray events beneficial

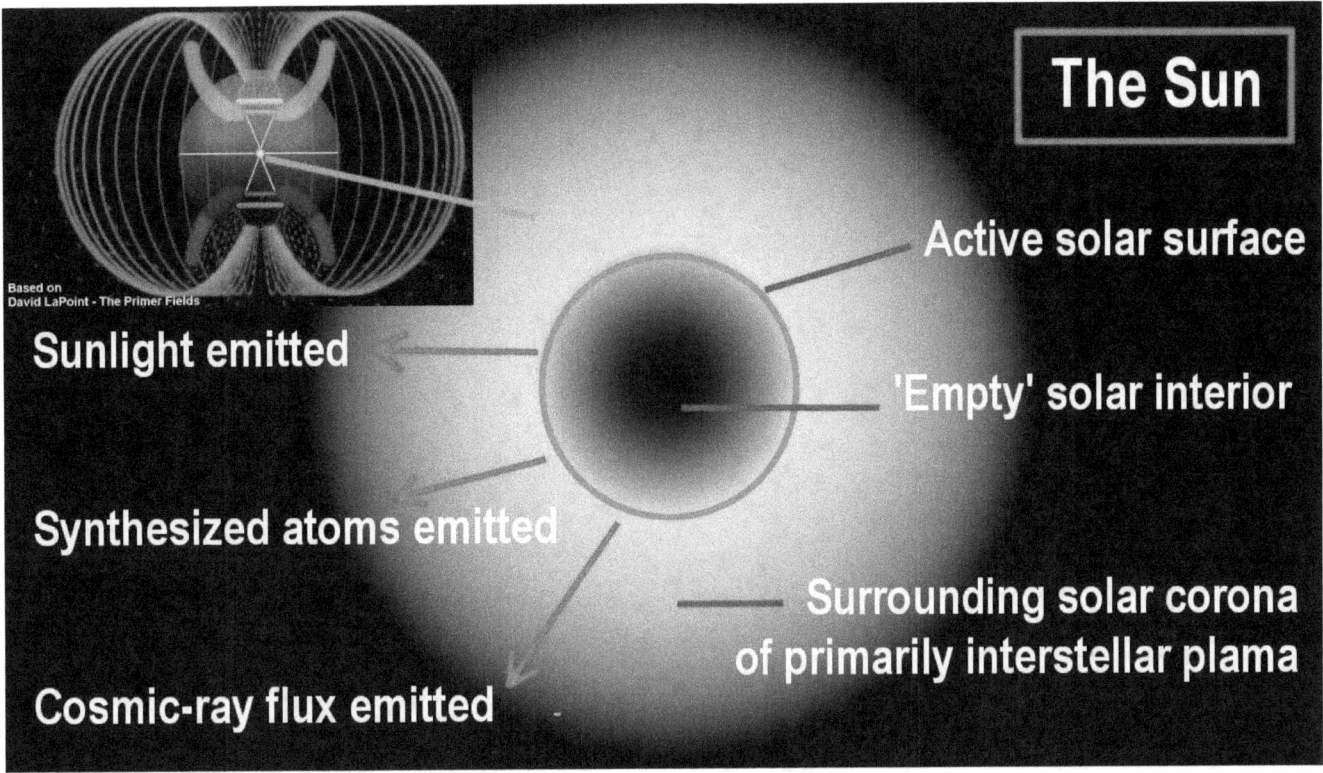

The weakening environment for the Sun enables larger volumes of Solar Cosmic-ray Flux to be emitted, which doesn't get trapped in the less-dense sphere of plasma around it. The resulting increase in cosmic-ray events is felt on Earth, including by us in a beneficial way.

Cosmic-ray particles 'see' our biology as empty

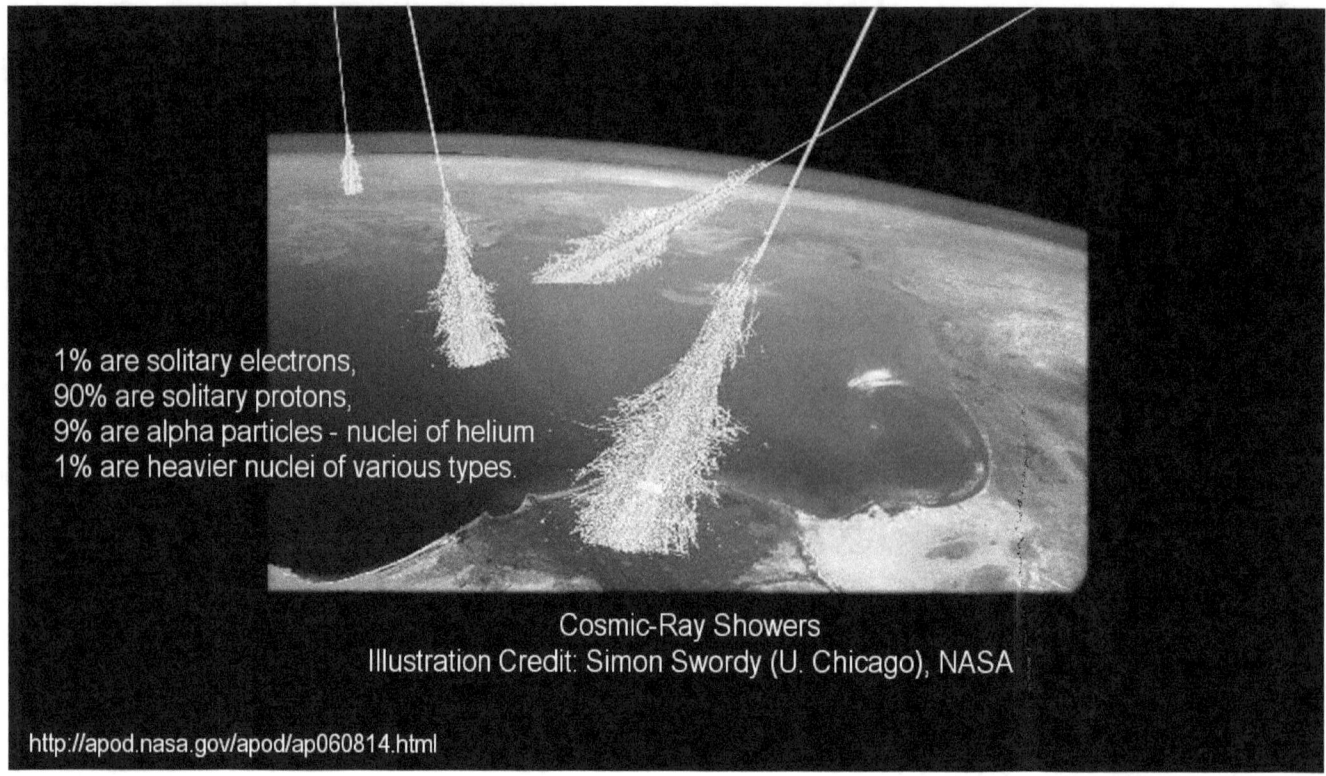

Cosmic rays are not continuous rays, like rays of light, but are single events of fast moving electrically charged particles, primarily protons. These particles are 100,000 times smaller than the atoms of our bodies. Atoms are constructs of plasma particles that are bound together into structures that comprise largely empty space. Cosmic-ray particles 'see' our biology as empty space and flow right through us without colliding with anything.

Electric currents by magnetic induction

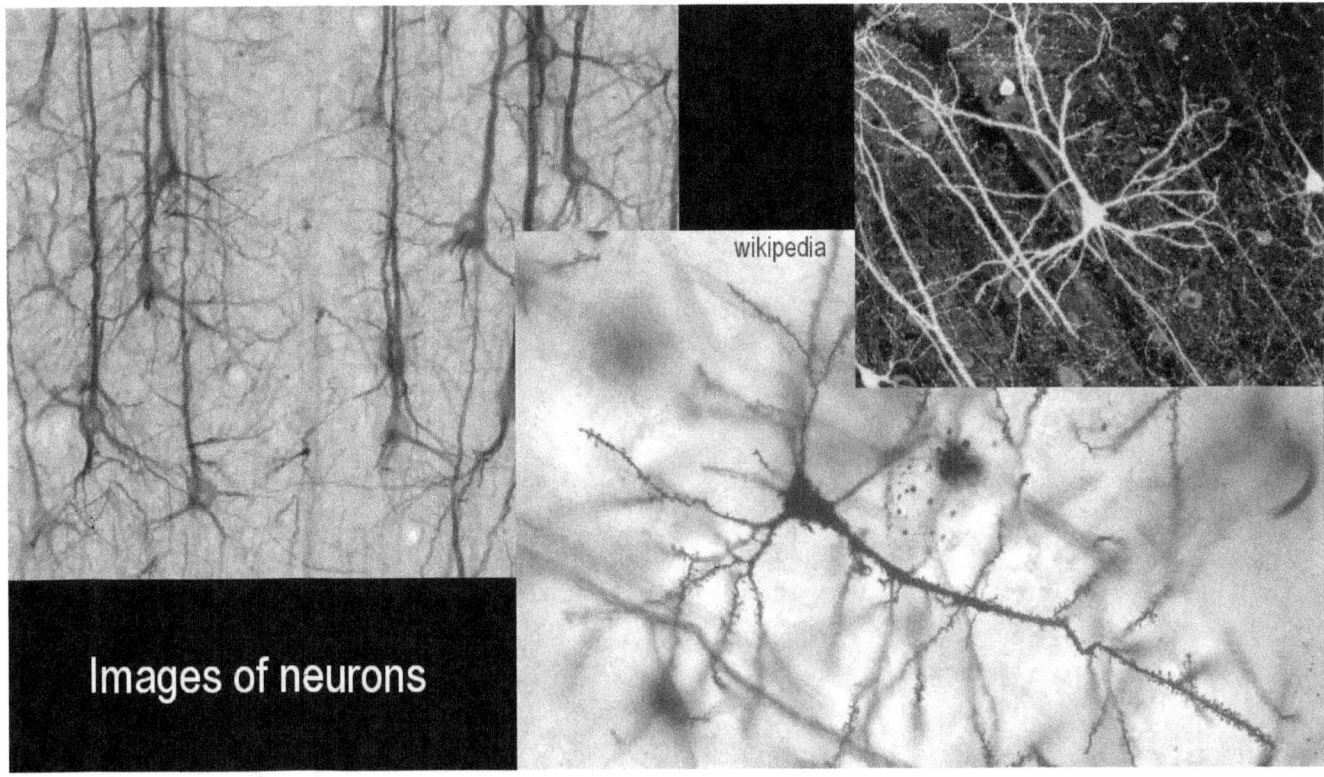

Images of neurons

However, their movement through our body leaves in the wake electric currents by magnetic induction.

Electric currents beneficial

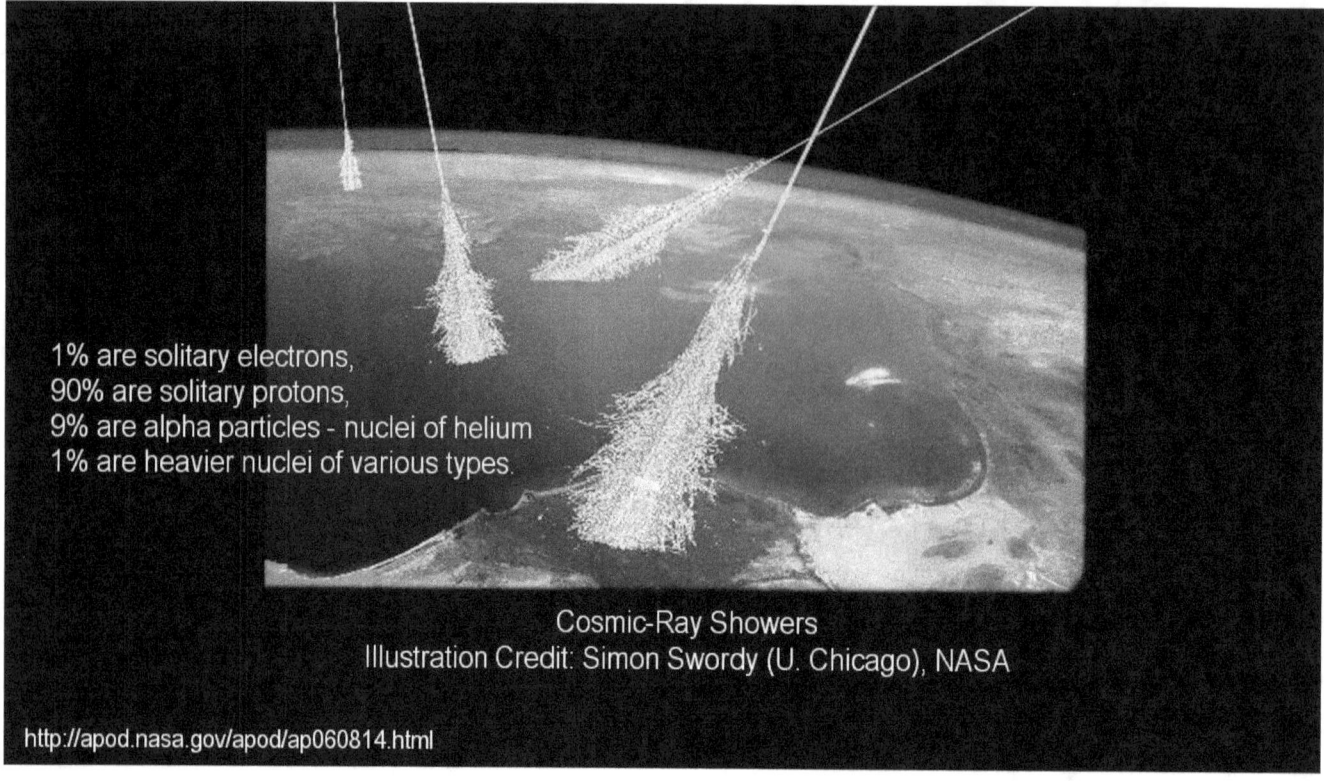

The electric currents appear to be beneficial for our complex neurological system that operates electrically to a large degree.

The greatest cultural advances

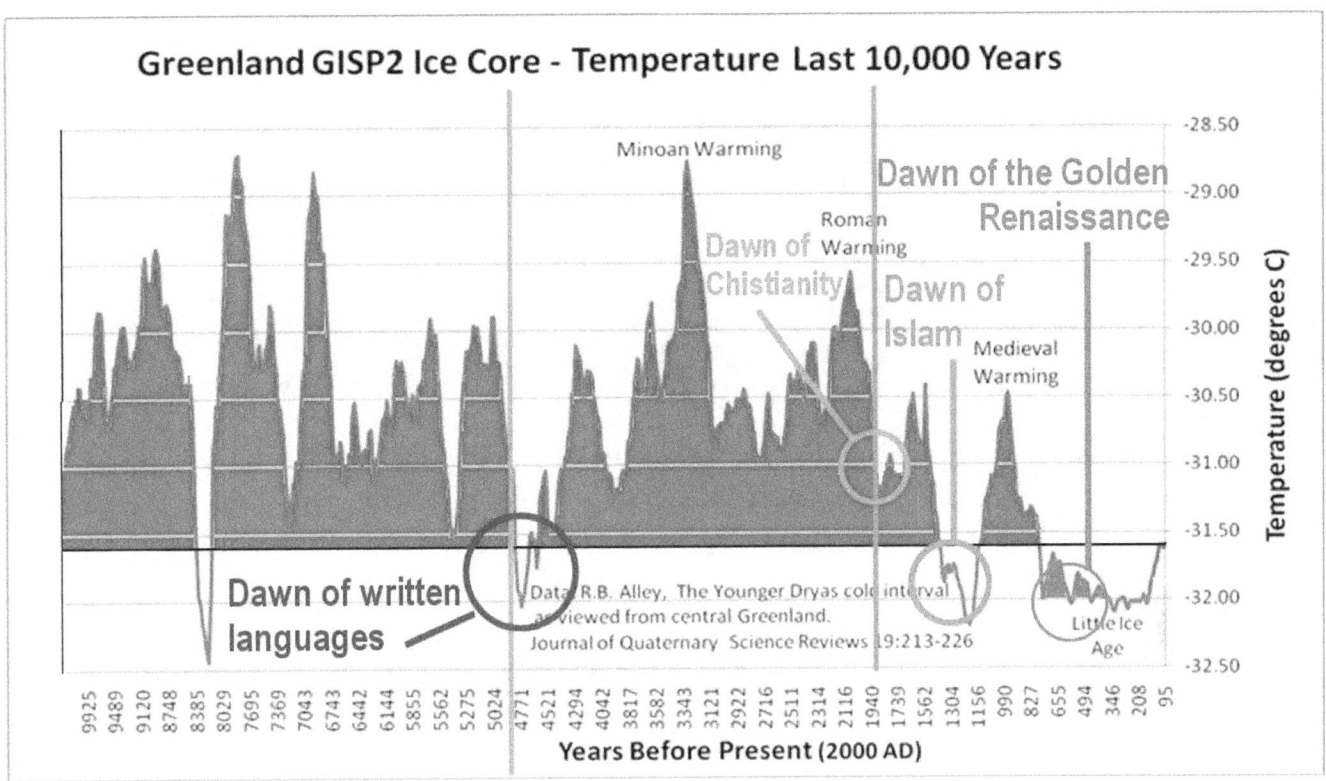

The evidence is surprising. History tells us that the greatest cultural advances in civilization were wrought during times of cold climates, which are typically times of larger volumes of cosmic-ray events occurring.

Lean times of cosmic-ray flux, times of war

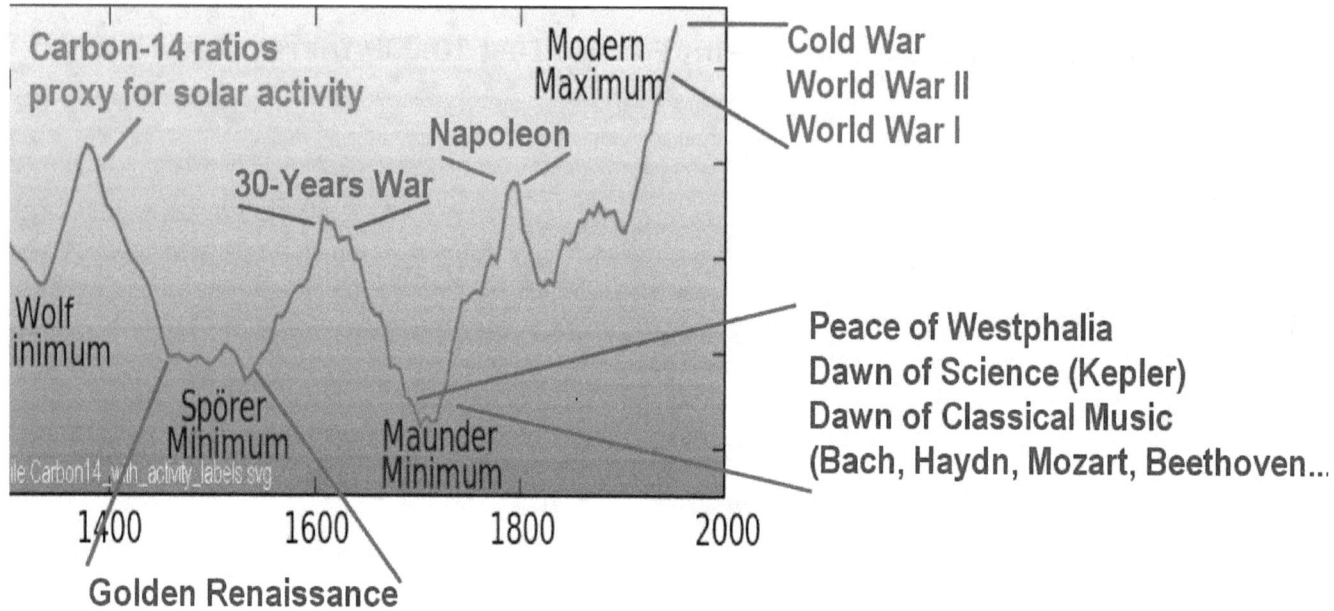

War and Renaissance

Inversely, the times of warm periods, which were lean times of cosmic-ray flux, where the times of the horrors of war.

I am not saying with this that the destiny of humanity is out of its hands. I am merely suggesting that a cognitively starving society, has a lesser chance to develop itself, than a well nourished society.

I am bringing this point up to illustrate that we actually do have a good chance that the needed 6,000 new cities, afloat in the tropics will be built, as we are on the fast track to ever-increasing volumes of cosmic-ray flux impacting the Earth, and us with it.

With the solar system getting weaker

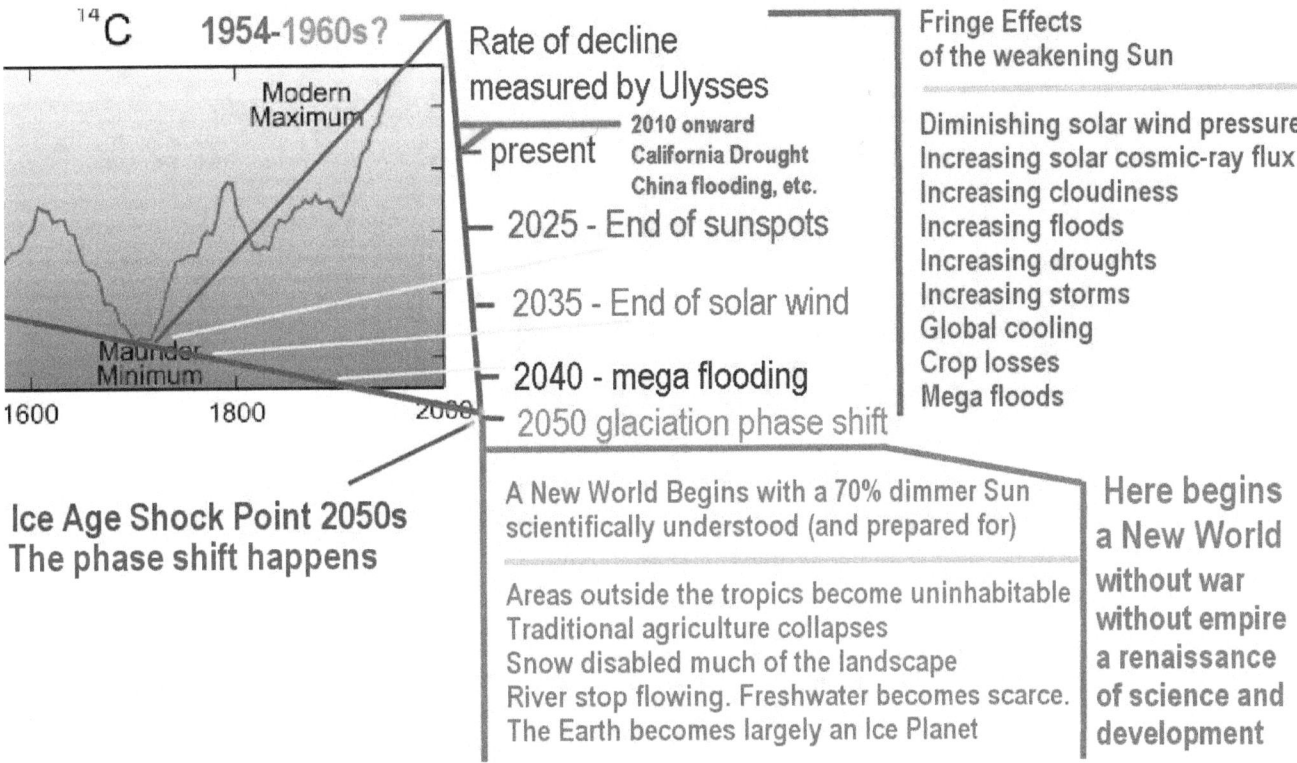

Ice Age Shock Point 2050s
The phase shift happens

With the solar system getting weaker at a rate of 30% per decade, we are way past the cosmic-ray starvation level in which the big wars happened.

The age of the wars appears to be over

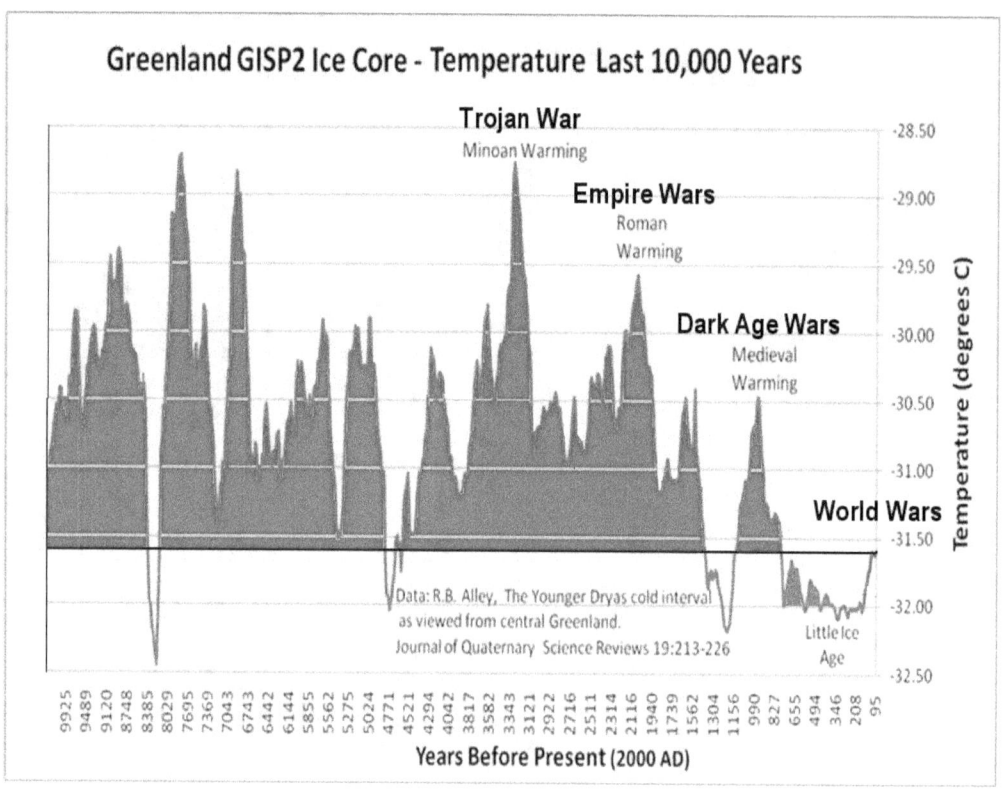

The age of the wars appears to be over. While the masters of empire are still rolling their war drums furiously, nothing will likely come of it. The period. of cognitive starvation has ended. A growing sense of peace and humanity is unfolding, faintly as this may be.

The greatest challenge in the history of humanity

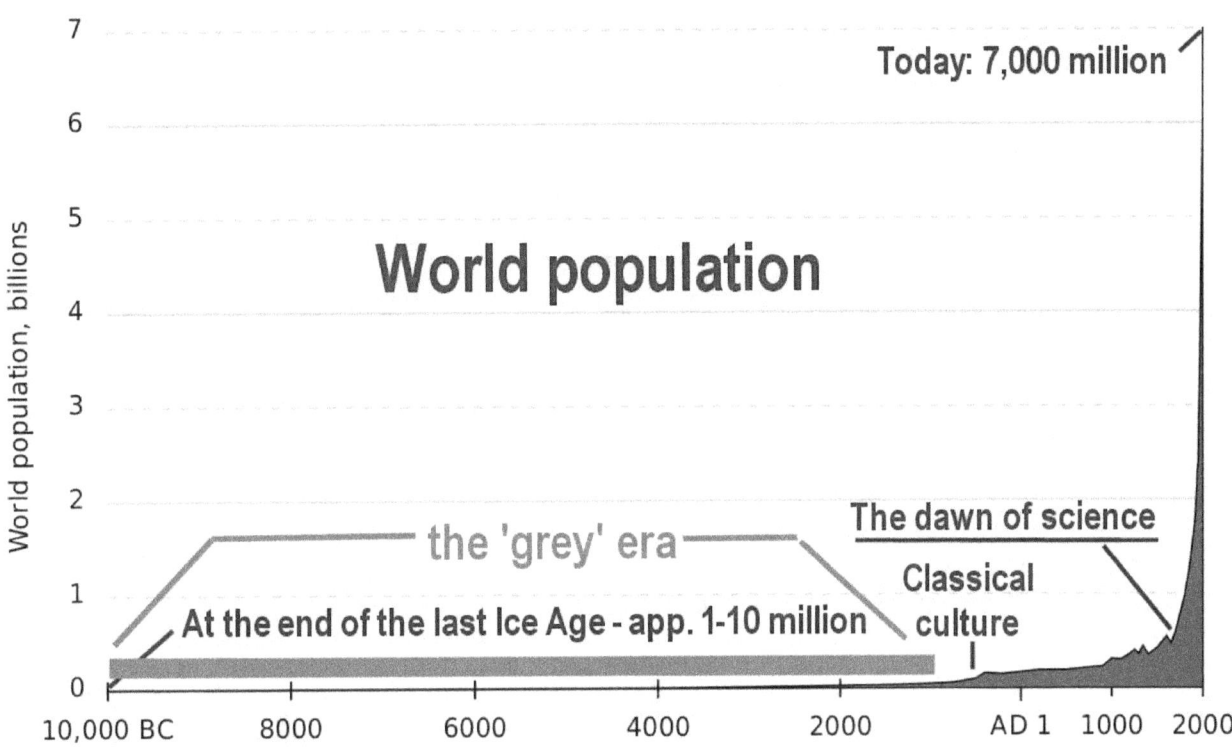

And still, another aspect of the Ice Age dynamics comes into view in the long historic sweep, when we open a window back to the last Ice Age and beyond. Aspects can be found there that are immediately relevant to the Ice Age Challenge of our modern time. The resulting view takes us far back into a history that is largely unknown.

But that's another story for another major exploration, and a necessary one, as the indifference in society in modern time goes very deep, is iron strong, standing in opposition to the greatest challenge in the history of humanity, which is the Ice Age Challenge to society to raise itself out of the easy chair and create itself a new renaissance world with technological infrastructures that the Ice Age glaciation cannot touch, and which has the potential to far supersede the present stage, with greater freedom, abundance, peace, security, beauty, and happiness.

The End

➢ More from the author:

14 Libraries of books and video productions

Novels on Universal Love, the greatest principle in civilization - **14 major novels**

Flight Without Limits (science fiction)

Brighter than the Sun (nuclear war avoidance?)

A series of twelve novels: **The Lodging for the Rose**
exploring the Principle of Universal Love

Book 1 - **Discovering Love**

Book 2 - **The Ice Age Challenge**

Book 3 - **Roses at Dawn in an Ice Age World**

Book 4 - **Winning Without Victory**

Book 5 - **Seascapes and Sand**

Book 6 - **The Flat Earth Society**

Book 7 - **Glass Barriers**

Book 8 - **Coffee Sex and Biscuits**

Book 9 - **Endless Horizons**

Book 10 - **Angels of Sex in Queensland**

Book 11 - **Sword of Aquarius**

Book 12 - **Lu Mountain**

The Sex and Sacrament Project - exploration stories from my novels - **11 books**

The Son of God

Impotence and Power

Self-Love and the Golden Hijab

Erica's Flower Garden

Helen a Healer

Brilliance of a Night

Gem of the Universe

The Sound of a Bird Woke Me

Between Ice and Spirit

Anton of Grace

Goodness of Living

The Kaleidoscope Project - mixed media of stories from my novels
- videos, PDF, audio

Discovering Infinity - developing history - 13 major research books:
A Research Book Series focused on scientific and spiritual development

Volume ii (Introduction) **Roots in Universal History** (Focus on Reality)

Volume 1A **The Disintegration of the World's Financial System** (Focus on Truth)

Volume 1B **Crimes Against Humanity** (Life Denied)

Volume 2A **Science and Christian Healing** (History as Truth)

Volume 2B **The Lord of the Rings' Metaphors**

Volume 3A **Universal Divine Science: Spiritual Pedagogical** (Structure for Discovery and Scientific Development - The Scientific Process to Know the Truth)

Volume 3B **Science and Health with Key to the Scriptures in Divine Science**

Volume 3C **Bible Lessons in Divine Science - 1898**

Volume 3D **Living in the Sublime**

Volume 4 **Light Piercing the Heart of Darkness** (The Demands of Truth and Justice)

Volume 5 **Scientific Government and Self-Government** (Platform for Freedom)

Volume 6A **The Infinite Nature of Man** (The Fourth Dimension of Spirit)

Volume 6B **Leadership** (The Spiritual Dimension of Leadership)

Cool Science of Kids - Illustrated Science - **interactive, videos, and 20 books**

War, Economics, and Nuclear War - scientific exploration - **10 videos**

Civilization - series focused on humanity - **10 videos**

Global Warming Doctrine - science videos - **12 videos**

Freshwater and Energy - science videos - **7 videos**

Christian Science explorations - **16 videos**

Books by Mary Baker Eddy - Christian Science - **16 on-line books**

Books by Rolf Witzsche on Christian Science - **9 Books**

The Giant PDF Library all transcripts of videos in PDF form

For links, please see: http://www.ice-age-ahead-iaa.ca

The projects are designed to draw the riches of our humanity into the foreground **towards a New Renaissance**, in order that their light may out-shine the systems of empire that are erroneously accepted, including the follies of war, terror, looting, economic destruction, science-perversion, and policies for depopulation.
Rolf A.F. Witzsche

www.ingramcontent.com/pod-product-compliance
Lightning Source LLC
Chambersburg PA
CBHW062353220526
45472CB00008B/1787